Mathmagical Moments

by Dr. Sue Heckler
& Dr. Christine Weber

Pieces of Learning

Cover design by Pat Bleidorn

©1995 Pieces of Learning
1990 Market Rd
Marion IL 62959
pol@dnaco.net
www.piecesoflearning.com

CLC0163
ISBN 1-880505-14-2
Printing No. 987654
Printed in the U.S.A.

Dedication

To our dads, John L. Johnson and Joseph Kazak, former teachers and administrators, who served as our role models. They provided the guidance for us to pursue what we have come to know as our greatest challenge — inspiring teachers to help their students reach their fullest potential.

Acknowledgements

To our publishers, Kathy and Nancy, who provided the spark for this project.
To Kathy Bleidorn for her math editing skills.
To teachers who reviewed the manuscript and provided lots of positive feedback.

About the Authors

Sue Heckler is a coordinator of gifted programs for Clark County Board of Education in Springfield, Ohio. She is a former secondary English teacher, media specialist and teacher of gifted K-8. Sue has been an adjunct professor at Wright State University in Dayton, Ohio, where she has taught media and computer skills and gifted education classes. She presents staff development workshops in the use of technology in the classroom, writing interdisciplinary curriculum, and developing thinking skills through content.

Dr. Christine Weber is a coordinator of gifted programs in Vandalia, Ohio, and former elementary classroom teacher. She served as a consultant in Indiana, Texas, Kentucky, and Ohio. Her academic credits include a Ph.D. in Curriculum and Instruction from Texas A & M University. Christine has conducted staff development workshops in developing critical and creative thinkers, integrating the curricula, and differentiating the curriculum to meet the needs of all students.

About this book

This book gives student practice for the strands as outlined by the NCTM standards.

Patterns
Problem Solving
Algebra
Measurement
Estimation
Probability

for 2nd-5th grade math students

A variety of activities are provided for each strand to enhance and challenge students' reasoning abilities. Reading level does *not* correspond to math level. **In fact, reading to the students is encouraged.** Then give students copies of the day's problem to solve. Most of the math word problems are **MULTI-STEP problems** so listening to them provides practice in memory skills.

This book also develops critical and creative thinking skills through interdisciplinary extensions which are presented in *ITALICS*.

Most importantly, when the teacher reads the daily social studies, science, and language arts trivia to the student, followed by the asking of a mathematical question, students practice two valuable skills —

MENTAL MATH

and important to all other subject areas
and
life long learning and living. . .

LISTENING SKILLS

How to Use this Book to . . .

☛ develop LISTENING SKILLS. Involve students in PROBLEM SOLVING
by instructing them how to listen to solve the story problems by
(1) listening for the question or the unknown and
(2) listening for details.
Students need to self-reflect about what they heard using pencil and paper.
Allow students to confer with another student or students to verify the
information before solving the problem.

☛ encourage JOURNAL WRITING. Use the extended activities in italics as
journal prompts. Involve students in the self evaluation of the problem solving
process using lead ins such as "What was puzzling about this?" — "How do I
go about solving this?" — "What other strategies could I use to solve this
problem?" — "How would -------- solve this?"

☛ MOTIVATE students by stimulating students' interests in a variety of
topics. Use the daily scenarios to encourage students to find more about
— research — people, places, and things.

☛ involve FAMILY members in what the student is learning. Share extension
activities in a newsletter or correspondence to parents to encourage extension
of topics and events that celebrate our lives.

☛ CHALLENGE students. Select the daily "challenge" problems or the
monthly Challenge for those students who have demonstrated advanced
mathematical reasoning skills or for those students who have exhibited an
interest in solving more complex mathematical problems. Challenge problems
can be used as part of authentic assessment.

☛ provide activities appropriate for AUTHENTIC
ASSESSMENT and student portfolios. Choose and allow your
students to choose activities from daily or challenge problems
and interdisciplinary extensions that reflect the
DEVELOPMENT OF THEIR REASONING
PROCESSES.

DAILY PROBLEMS FOR USE WITH AUTHENTIC ASSESSMENT

The following suggestions provide ways in which the daily problems can be used in the development of student portfolios for the purpose of authentic assessment of mathematical reasoning abilities.

September 22   In order to help students identify different flavor **combinations**, provide precut colored scoops of ice cream for students to manipulate and solve the problem. The finished product could be a chart or poster advertising the different cone combinations for sale in an ice cream parlor.

October 11   Have students present their survey **results in a variety of ways**. Different products that reflect the data collected may include a newspaper story, letter to the pizza company president, chart, graph, advertisement, or paragraph.

November 18   Provide opportunities for students to become better problem solvers by encouraging them to **solve problems in many ways**. Provide opportunities to use different problem solving strategies such as role playing the problem, making a model, or making a drawing or diagram.

December 22   Help students to understand the **relationship of objects** according to weight. Encourage the use of graphic organizers such as a ladder, pyramid or bar graph to show the relative weights of animals, plants or other objects.

January 13   Emphasize the use of **different combinations** of paper and coins to purchase an item. Encourage students to practice combinations by setting different criteria such as using all paper money, all coins, one paper bill and 4 coins, 5 different coins, etc.

February 23   Provide lots of experiences for students to investigate the **various meanings** of "one-half." For example, how could you describe one half of the following—my math textbook, ten cookies, a glass of water, an ice cube, a puzzle, friendship, etc. Have students create a poem that describes one half of an object or idea.

March 9   Allow children to explore the concepts of beginning, middle and end as related to the **measurement** of time. Students could develop time lines reflecting the events of a certain day, the life of a pet or other animal, a period of time, or a favorite book.

April 14   Expand students' knowledge of **geometric patterns** by brainstorming geometric shapes, either man made or in nature, in order to construct a collage or three dimensional models which have drawings or pictures on each surface.

May 18   Involve students in the discovery of **patterns** by systematically enlarging and decreasing the length and width of rectangles. Have students identify the pattern. Encourage students to research a variety of patterns and then create their own book of patterns.

INTERDISCIPLINARY EXTENSIONS

The following suggestions provide ways to use interdisciplinary connections in the development of student portfolios for authentic assessment of critical and creative thinking skills.

September 2
Provide opportunities for students to compare and contrast nursery rhyme characters using a graphic organizer such as a Venn diagram or a chart.

October 19
Provide an experience for students to develop a personal analogy by asking them which shape they are most like. Encourage students to explore the similarities of their attributes and the attributes of their chosen shapes. Have students complete their analogies in the form of a poem, illustration or paragraph.

November 12
Given the prompt "*Would you like to live in the frozen land of Antarctica? Why or why not?*" encourage students to develop a response paragraph that includes the development of the main idea and supporting details.

December 13
Ask students to analyze in a journal the successes and failures of their experiments with dissolving lifesaver candies. Ask relevant questions that students can respond to in journals. Help children determine specific reasons why one method works better or best.

January 11
Help students picture the stages of development / progress of their ice cream and banana dessert by developing a sequence chart, a flow chart, and by writing directions for others to follow.

February 7
Develop creativity by allowing students to brainstorm responses to "What would the world be like without books?" Encourage lots of ideas (fluency), different kinds or types of ideas (flexibility) and unique or unusual ideas (originality). Provide opportunities for students to expand upon (elaboration) a response by developing an illustration, poem, story, or press release.

March 2
Guide students to develop criteria to evaluate their cartoon collages as a whole class. Have each student use the criteria to judge his/her own collage and a classmate's collage.

April 2
Model the writing of a diamante poem using the opposites *ugly* and *beautiful*. After brainstorming nouns, verbs, and adjectives, have the students create a diamante using opposites.

May 15
Have students choose one of the eight bears and research their bears in a non-fiction book AND an encyclopedia. Have them summarize the article and book in two separate paragraphs.

Edgar Rice Burroughs, "Tarzan's" inventor, was born in 1875.

Mr. Burroughs wrote 67 books including 23 stories about Tarzan and his ape friends. In the books, Tarzan, the son of a nobleman, is left in Africa and raised by a family of apes. As Tarzan grows up in the jungle, he has many adventures. **How many of Burrough's books were not about Tarzan?**

What are some names Tarzan could have called his ape friends?
How is Tarzan like a basketball player?
What would the forest floor look like to an ape in the trees?
What if Tarzan would have grown up with bears instead of apes?
Draw a picture of Tarzan and his friends in the forest.

Peter and Iona Opie collected nursery rhymes.

The Opie's edited a collection of over 500 nursery rhymes entitled ***The Oxford Dictionary of Nursery Rhymes***. The book, which took them six years to write, includes the history and the earliest versions of the rhymes. They also wrote ***The Oxford Nursery Rhyme Book*** which contains 800 nursery rhymes. **How many rhymes are contained in their two books?**

How is a nursery rhyme like a folk tale?
Choose your favorite nursery rhyme and write a diary from the viewpoint of the main character (for example, how did Jill feel about her experience on the hill?)
Read 2 or 3 nursery rhymes and compare and contrast the characters in them.
Write an editorial about "Tom, Tom the Piper's Son."

Viking II, a U.S. satellite, landed on Mars in 1976.

Before satellites were launched into space, scientists looked at the planets through telescopes. Viking I landed on Mars on July 20, 1976. Then Viking II landed. Both satellites sent pictures of Mars back to the United States. **If the earth is 93,000,000 miles from the sun and Mars is 141,500,000 miles from the sun, how many miles is Mars from earth if they were in a straight line — sun, earth, Mars?**

Would you like to take a space ride to Mars? Why or why not?
What would be different if we lived on Mars instead of earth?
Design a craft that could take you and four friends to Mars.

A record long go-cart journey was held in Canada.

A four-man team drove their go-cart 12,018 laps over a 24 hour period on September 4th and 5th in 1983. **If every 6 laps were one curvy mile in length, how many miles did they drive?**

List the different feelings you would have if you went on the 24 hour go-cart trip.
What would happen if the course was straight instead of in a circle?
Name things you can carry in a go-cart. List things that go fast. List things that have curves. How is a curve like a baseball diamond? Design a go-cart on paper. Select criteria to judge the design. Select the class's best designs based on the criteria. Award certificates to the best designers.

One of the longest tunnels in the world opened in Switzerland in 1980.

Some tunnels must go through hard rock and are usually blasted with explosives. Cutting machines called *moles* are used to dig through softer rock and soil. One of the longest tunnels in the world, St. Gotthard's Tunnel, connects Switzerland to the rest of Europe. The tunnel is 10 miles long and took 10 years to build. **Estimate how many miles were built each year.**

CHALLENGE If the tunnel cost $417,000,000, how much did it cost per mile?

Name things that tunnel. How is a tunnel like a bridge? Make a list of other important tunnels in the world. Locate these tunnels on a map. What would happen if there were no tunnels or bridges in the world? Build a clay model of a tunnel and label the parts.

St. Gotthard is one of the world's longest tunnels. What is the longest bridge?

The earliest bridges were fallen logs or vines that helped a person cross a stream. In order to cross wider bodies of water, different ways of building bridges developed. One type is the suspension bridge. It hangs on cables that are fastened to tall towers on the river bank. One suspension bridge is the Humber Estuary Bridge in England. It was completed in 1980 at a cost of about $185,000,000. Its towers are over 500 feet tall and the main bridge spans 4,626 feet. Drivers must pay a toll in order to cross the bridge. **If you crossed the bridge 20 times a month, and each time you paid $1.00, how much would you pay in bridge tolls in one year?**

List things that are tall. How would a bridge look to a fish swimming under it? An airplane flying over it? Compare and contrast tunnels and bridges in a Venn diagram. Draw two different types of bridges and label the parts.

Grandma Moses's birthday is today.

Anna Mary Moses started painting when she was 76 years old. Her paintings showed farm scenes that she remembered from her childhood. She painted 25 of her pictures after she was 100 years old. **If she was born in New York in 1860 and died in 1961, how old was she when she died?**

If you could talk to Grandma Moses, what would you ask her? Do you think she would like to watch television? Why or why not?
Draw or paint a picture of your home and yard like Grandma Moses did. Give it a title. Display your pictures in your classroom art gallery.

Bert Campaneris played every position on a baseball team in one game.

Baseball is considered the national game of the United States. In 1965, Mr. Campaneris played a different position each inning of one game. He pitched the 8th inning. **What was the total number of positions he played during one game?** Name the other positions he played in this game.

CHALLENGE A pitcher pitches 3 innings in the first game and 3 innings in the second game of a doubleheader. If he allows 3 runs in each game, what is his average run per inning?

Name the shape of a baseball field. How would the game be different if the field were a circle? a square?
How is a baseball like a fire engine?
Name other games where players use some kind of a ball.
Write a news story about Mr. Campaneris' unusual game.
Make rules for a new game using cotton balls. Play the game with a classmate and evaluate the game.

Colonel Sanders, founder of KFC ® , was born today.

Colonel Sanders created a recipe for cooking chicken that all his friends liked. He started a restaurant to sell his special chicken. **If you went to a *Kentucky Fried Chicken* restaurant and ordered a piece of white chicken for $1.21, a serving of mashed potatoes and gravy for $.69, and a medium drink for $.89, how much change would you get from a $5.00 bill?**

What other foods could Colonel Sanders sell in his "chicken" restaurant?
What adjectives could you use to describe Colonel Sanders' chicken?
Compare and contrast Colonel Sanders' menu with another fast food menu.
Create a new menu item for Colonel Sanders to sell in his restaurant.
Design a menu with items from Colonel Sanders' restaurant.

Charles Feltman introduced the first hot dog in a bun.

The original hot dogs were called frankfurters and probably came from Germany. In the United States, they are called hot dogs and are a favorite American food. In 1919, Mr. Feltman, a New York Coney Island baker, baked a bun to fit a hot dog and sold them from his bakery. **If he sold his hot dogs for $.30 per package of ten, how much did one hot dog cost?**

How much more do hot dogs cost today?
Where are all the places you can buy hot dogs? What other names could you call a "hot dog"?
How is a hot dog like a firecracker?
Draw a "hot dog" cartoon character.

Penny, a Buckinghamshire hen, laid 7 eggs in one day.

Penny, an English hen, broke a record by laying 20 eggs in one week and 7 in one day. **How many eggs did she have to lay the other days of the week to reach 20 eggs? If a hen lays 7 eggs one day, 6 eggs the next day, and 7 eggs the third day, what will she lay the fourth day if she continues this pattern?**

Make a list of questions an egg would ask an egg carton.
How is an egg like a chicken?
What are other names you might give Penny?
What can you do with eggs without breaking them?
Design a safe way to carry eggs home from the store.

A very high score for a soccer match occurred in a Scottish Cup match.

Soccer games usually have scores of less than 10 goals for each team. The score in the 1985 Scottish Cup game was 36-0. **What would the half time score be if the team scored one half plus 3 of its goals in the first half?** A soccer ball is slightly smaller than a basketball. One of the world's largest soccer balls was made in England. It measured 7 3/4" in diameter and weighed 80 pounds. **If a basketball weighs 3 pounds, how much more did the largest soccer ball weigh?** *Compare soccer and basketball using a Venn Diagram.*

CHALLENGE Collect string or yarn to make a ball. Measure the string before you roll it into the ball. How much string does it take to make a ball 1 inch in diameter? Estimate how much string it would take to make a ball 1 foot in diameter.

The highest "shade" temperature was recorded in 1922.

How hot is hot? In Al' Aziziyah, Libya, the temperature was measured at 136.4° F in the shade. **If the average summer temperature is 102.2° how much hotter was it in Libya?**

Name things with degrees.
List things that make shade.
Draw a picture of things that might be happening on a hot, sunny day.
Read Robert Louis Stevenson's poem "The Shadow." Write your own "shadow" poem.
Compare and contrast your shadow with a friend's shadow.
How can you change your shadow's shape?

John Steptoe, black children's author, was born in 1950.

Mr. Steptoe was a teacher and illustrator. He wrote his first book, **Stevie**, when he was 17 years old. **In what year did he write the book?**

Read several books by John Steptoe. Compare and contrast the characters in the books and write a paragraph about them.
Choose your favorite Steptoe book and give a book talk to encourage others to read it.
Create a diorama of your favorite scene from this book.

Tomie De Paola, children's author and illustrator, was born in Connecticut.

When Tomie was in elementary school, he told his teachers he wanted to write and illustrate picture books when he grew up. His book titled **Strega Nona** was selected as a Newberry Honor book. Two other books are titled **Pancakes for Breakfast** and **Popcorn**. **If a recipe for pancakes calls for 2 cups of flour for 12 pancakes, how many cups of flour would you need to make 24 pancakes?**

Make a list of questions pancakes would ask a bottle of syrup.
How are pancakes like popcorn? Invent a menu for a healthy breakfast (include all the food groups).
If a recipe for pancakes calls for 2 cups of flour for 12 pancakes, how many cups of flour would you need to make 1 pancake for each of your classmates?

The Mayflower set sail.

The *Mayflower* left Plymouth, England, in 1620, and finally arrived in Plymouth, Massachusetts. The *Mayflower* had to turn back twice because the ship sailing with them, the *Speedwell*, had leaks. The *Mayflower's* journey took 66 days. The *Speedwell* was unable to make the same voyage. The settlers from the *Mayflower* landed near Plymouth Rock. Thirty-five of the 102 passengers were Pilgrims from Holland. **How many of the passengers were not from Holland?**

CHALLENGE What percentage of those aboard the boat were Pilgrims from Holland?

Who was the youngest Pilgrim? What other rocks can you name? If you could be any kind of rock, which one would you choose to be? Write about one day in the life of the rock you have chosen.

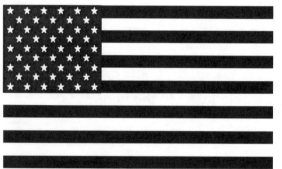

Celebrate Citizenship Day.

Citizenship Day has been celebrated since 1952. Speeches and pageants recognize and celebrate the privileges and responsibilities of being American citizens. The American Constitution was also signed on this date in 1787. **If one class of 26 creates buttons to wear for themselves and two other classes with 22 and 30, how many buttons will they make?**

Create several mottos for Citizenship Day.

_____

_____

_____

_____

In the space below create buttons for one or two of the mottos.

Tom Thumb lost a race with a horse.

The famous race between a horse and a steam locomotive, *Tom Thumb*, took place in 1830. The builder of the *Tom Thumb* was trying to prove that the locomotive was a better way to carry goods than a horse. As children and adults we collect toy trains and model trains that are built to scale. **What does "scale" mean?**
Popular scales of trains include 0 = 1/48, HO = 1/87, and N = 1/160.

CHALLENGE *Which is the largest model? Smallest model?*

Who won the race? Why?
Why were locomotives sometimes called "Iron Horses?"
What are some other uses for scales?
List things that rhyme with scales. Create a poem using those words.
Make a time line about the history of the train.

September 19

Mickey Mouse appeared in "Steamboat Willie" in 1928.

"Steamboat Willie" was the first cartoon with a soundtrack. Walt Disney was Mickey's voice. Animals like Mickey Mouse were easier to animate than people. Why? Mickey was drawn using different sized circles. Draw your favorite animal using only circles.

Times Square in New York City is a famous square. The Loop in Chicago is a famous loop. List other famous squares and circles. Learn to draw an animal using geometric shapes.

September 20

George Simpson received a patent for the electric range.

Simpson's patent, obtained in 1859, was for a cooking range which made heat by passing electricity through coils. **Name tools of measurement used in cooking.**

In what ways did the electric range change the lives of people?
Change a favorite recipe to metrics. Exchange metric recipes with a friend.
With help from an adult, make this recipe or create your own recipe using only metric measurement. Bring your food to class. Write a poem to introduce the foods found in your recipe.

Herbert Wells was born.

Wells, the author of science fiction books such as *The Invisible Man, The Time Machine,* and *The War of the Worlds* was born in 1866. He was 32 when he wrote *The War of the Worlds.* **In what year did he write this story?**

Many of his books were made into films.
Which do you like better— to read the book or watch the film? Explain why.
List things that are invisible.
Using your imagination, visit the past or future in your own time machine.
Make a diorama of what you see.

September 22

The first ice cream cone was patented.

The patent for the ice cream cone was issued to Italo Marchiony in 1903. The cone was first made of paper and later made of pastry. Sue and Joe love ice cream cones. **If Sue and Joe bought 3 flavors — vanilla, chocolate, and strawberry — how many different double dip combinations are possible?**

CHALLENGE If you bought a double dip cone at the store, a vanilla scoop may cost you $.40 and the other flavors (chocolate and strawberry) may cost $.45. What is the probability that the double dip cone will cost $.90?

What name would you give to a new kind of ice cream that does not melt? Write a jingle including the new name.

September 23

The first airmail pilot was hired in 1911.

Earl Lewis Ovington was known as "air mail pilot number one." In his plane *Dragonfly* he delivered mail from Garden City, New Jersey, to Minneola, Long Island, a distance of six miles. The first mail consisted

of 640 letters and 1280 postcards. **How many more postcards were carried than letters?**

CHALLENGE What percentage of the mail carried was letters?

Trace Ovington's flight on a map. Why was "Dragonfly" a good name for this airplane? Compare and contrast a letter with a postcard. Which would you rather receive — a letter or a postcard? Why?

The first National Monument was Devil's Tower.

Devil's Tower is a volcanic rock 865 feet tall. It is located in Wyoming. President Roosevelt signed a bill which made 1153 acres surrounding Devil's Tower a part of this monument. Gannett Peak, the highest point in Wyoming, is 13,804 feet tall. **How much higher is it than the Devil's Tower Monument?** Round your answer to the nearest thousand.

There are 79 National Monuments. Which monument would you like to visit?
Survey how many students in the class have visited a National Monument. Make a chart of the results.
How would Devil's Tower look to a bumblebee?
A passenger in a helicopter?
An astronaut in a spaceship?
Illustrate one of the views.

The first telephone conversation over the ocean took place.

In 1956, a telephone call between New York City and London occurred. This first commercial call cost $12 for every three minutes. **If you spoke to a friend in England for nine minutes using this rate, how much would your conversation cost?** How many miles did this first telephone conversation cross?

List different ways people communicate.
Develop a new way to communicate with a friend.
Develop a message in code, and ask a friend to figure out what it says.
Invent a new way to give directions from your house to the library.
Write a riddle or pun about the telephone.

Shamu was born today.

Shamu was the first killer whale to survive that was born in captivity. One of the places Shamu lives is at Sea World in Texas. A killer whale, also known as a toothed whale, can have up to 48 teeth. **How many more teeth does this whale have than you?**

Do all whales have teeth?

_____

What does it mean to have "a whale of a good time?" Give an example.

_____

_____

_____

_____

What other sayings have animal names in them?

_____

_____

_____

Create your own sayings using animals. Illustrate their meanings.

_____

_____

_____

_____

Write a legend describing how one of your sayings came about.

16

A patent was obtained for a book of matches.

Joshua Pusey of Lima, Ohio, received his patent in 1892. His matches were made in books of 50 matches. The matches were struck from the inside cover of the book. This was very dangerous. The other matches in the book could easily catch fire. It was several years later when books of matches were made to be more safe and usable. **If campers took 5 books of Pusey's matches and used 16 matches to light fires, how many would be left?**

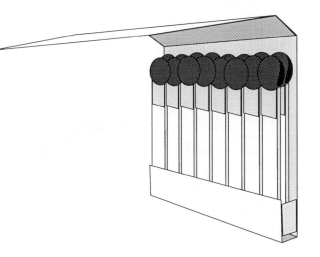

What would happen if there were no matches in the world?
How is a match like a message?
Match is the answer. What could the questions be?
List things that match.

San Diego Bay was discovered.

Rodriguez Cabrillo arrived in California in 1542. He and his crew discovered San Diego Bay. Rodriguez died from a fall the following year. A monument was built in his name 370 years after his death. **In what year was the monument built?**

Was it possible for Cabrillo to use a magnetic compass on his voyage?

What might have been Cabrillo's first words when he arrived in California?

Make a journal entry for September 28, 1542, by one of the crew members.

Compare and contrast a monument with a diary.

The heaviest man on record was born in 1941.

In 1963, the heaviest man weighed 392 pounds. In 1966, he weighed 700 pounds. In 1976, he weighed 975 pounds. He died on September 10, 1983, weighing 98 pounds more than he did in 1966. **How much weight did he lose from 1976 to 1983?**

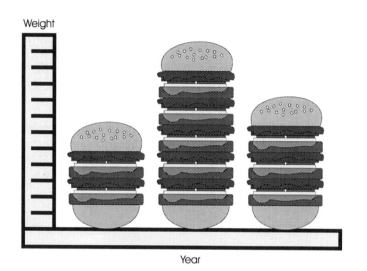

Make a bar graph showing his weight from 1963 until his death.
Make a chart with pictures found in newspapers and magazines of items that weigh between one and 10 pounds. Arrange the items in descending order of weight.
If you could break a world record, what would it be?

The first gerenuk is born in the United States.

This reddish brown gazelle-like antelope has a long neck. It is originally from eastern Africa. The first gerenuk born in the United States was born in New York City in 1963. **If a new gerenuk was born every other year since then, how many would there be?**

Gerenuk is the answer. What could the questions be?

As a class, make a mindmap about gerenuks.

Is the gerenuk considered an endangered species? How could you find out?

18

September Challenge

people	places	things	numbers
Joe	school	desk	7
Steve	playground	ball	9
Jessica	backyard	swing	3

Using any or all of the people, places, things, and numbers above, write 4 word problems — one for addition, one for subtraction, one for multiplication, and one for division.

1. _____

_____

2. _____

_____

3. _____

_____

4. _____

_____

Pele, a great Brazilian soccer player, retired in 1977.

Edson Arantes do Nascimento was the son of a professional soccer player. When Edson was eight years old, his friends gave him the nickname, Pele. At 15, he became a professional player for Santos Soccer Club in Brazil. In 1959, he scored a record 127 goals in one year. In October, 1974, Pele retired from playing soccer in Brazil. Then he played soccer in the United States. Pele helped the game of soccer become popular in this country. At the time of his first retirement, he had played in 1,363 games and scored 1,281 goals. **Estimate how many goals per game he scored.**

How is a soccer ball like a book?
Name other kinds of goals.
Choose two goals for your personal growth to work toward this school year. Write them in a notebook. At the end of the year, evaluate how well you achieved your goals.

October 2

"Peanuts" was first published as a cartoon in 1950.

Charles Schultz is the creator of the *"Peanuts"* cartoon. *Charlie Brown* and *Snoopy* are characters in the cartoon strip. *Charlie Brown* is the underdog. *Snoopy* is a dog who acts like a human being. The *"Peanuts"* cartoon is published in about 1800 newspapers in the United States and Canada. **If daily newspapers sell for $.30 a day and the Sunday paper costs $.50, how much would it cost to receive the newspaper for a 28 day month that has 4 Sundays?**

How is a cartoon like a piano? What is an underdog? How does it feel when you are the underdog? What would happen if Snoopy were a cat? Compare and contrast Snoopy with Garfield. Create a cartoon animal that acts like a human being.

October 3   In 1964, a very long golf game ended.

Floyd Rood played golf across the entire United States. He began at the Pacific Ocean and hit the golf ball until it reached the Atlantic Ocean. During his golf game, he covered 3,397.7 miles and hit the ball 114,737 times. He lost over 3500 golf balls. **If he started on September 14, 1963, how many days did the golf game last?**

How is a golf ball like a kangaroo?
Where are some of the places Mr. Rood might have lost his golf balls?
(Use a topography map to find places the golf balls might have been lost).
What might his lost golf ball say to the person who finds it?
How would you feel at the end of this long golf game?

October 4

Truck drivers drive long hours on the highways. They have a set of words they use to talk to each other using a Citizen's Band Radio or CB. Each CBer has a nickname called a *"handle."* The truckers talk back and forth from truck to truck using their handles and code words. When they are finished they sign off with *"ten-four."* Today is *"ten-four"* day because it is the 4th day of the 10th month. **If a truck driver drove 500 miles in one day, how long would it take him to complete a 1,500 mile trip?**

Name things with handles.
How would you feel if you traveled many days at a time away from home, family and friends?
What would happen if trucks ran on railroad tracks?
Create code words and meanings for words you use each day in school (example: recess, silent reading, lunch).
Design a code book with these words.

October 5

In 1974, a pogo stick jumping record was set at 6 hours and 6 minutes.

A person using a pogo stick and kangaroos are great jumpers. Kangaroos actually spring along using their powerful back legs for bounce and their tails for balance. Douglas Zeigler continuously jumped up and down on a pogo stick for over 6 hours. He jumped over 40,000 times. **Estimate how many times an hour he jumped.**

Name things that jump.
How is a pogo stick like a bird?
What if you could jump like a kangaroo?
Would you rather be a kangaroo or a koala bear? Why?
Draw a picture of a person on a pogo stick racing a kangaroo.

October 6

George Westinghouse, American inventor, was born in 1846.

George worked in his father's machine shop as a boy. He was always inventing something. He received a patent for his first successful invention, an air brake, in 1869. This special brake was used to make train engines easier to stop. Before he died in 1914, he had over 400 patents. **How old was Mr. Westinghouse when he received a patent for the air brake?**

Name things that you can break. What other machines might use an air brake? How would a speeding train look to an airplane? Compare and contrast a bicycle brake with Mr. Westinghouse's air brake. Write a letter to Mr. Westinghouse asking for his advice on something you have invented.

October 7

A photograph of the hidden side of the moon was taken in 1959.

For the first time scientists could actually see how the "other" side of the moon looked. The pictures of the moon showed thousands of craters made by meteorites. The U.S.S.R. space ship, Luna III, took the photographs at 6:30 a.m. The space ship was 43,750 miles from the moon. The photographs were transmitted 292,000 miles back to earth. **If it took 4 hours and 10 minutes to reach earth, what time did scientists on the earth see the photo for the first time?**

Draw a picture showing how you think the moon's other side looks. Name things that are hidden. How is the moon like a panda bear? Mindmap the topic "moon."

October 8

The great Chicago fire disaster occurred in 1871.

According to stories about the fire, Mrs. O'Leary's cow kicked over a lantern in her barn. This started the fire which destroyed much of the city of Chicago. Before the fire, many buildings were made of wood. When people started to rebuild the city, they wanted the buildings to be made of a material that wouldn't burn easily. By 1875, the world's first metal frame skyscraper was built in Chicago. **How many years after the fire was the skyscraper built?**

CHALLENGE If Mrs. O'Leary's barn was 24 feet wide and 38 feet long, what was the area in square feet?

List things that are the color of fire. Name ways to prevent fires. In what ways does fire help us? Design a fire safety poster. Create a slogan for the poster.

October 9

Today is Korean Alphabet Day.

The word *alphabet* comes from the Greek words *"alpha"* and *"beta"* which are the first two letters of the Greek alphabet. An alphabet is a collection of letters used to write words of a language. Different countries have different alphabets. In 1446, King Sejong the Great of Korea proclaimed a new alphabet for his people. **How many years has the Korean alphabet been used?**

Where does our alphabet in the United States come from? Learn to draw a letter from the Korean alphabet. Design a new way to write the letter "A." Make a class chart with unique words that begin with different letters of the alphabet. What if our alphabet only had 24 letters? Which letters would you omit? Why?

October 10

A 98 year old man finishes a marathon race.

The word *marathon* means *"something with endurance"* like a long distance foot race. Early marathon races were different lengths. In 1924, the marathon race was standardized at 26 miles, 385 yards. In 1976, ninety-eight year old Dimitrion Yardanidis completed a marathon in Athens, Greece. It took him 7 hours and 33 minutes. An 82 year old female runner completed her marathon in 7 hours and 58 minutes. **How much faster did Dimitrion run?**

List other things that could be marathons.
How is a marathon like a spaceship?
How would you feel if you ran a marathon race at age 98? Design a mini-marathon race around your school (26 yards, 385 feet or 26 feet, 385 inches). Time your classmates to see who is the fastest mini-marathoner.

One large pizza was eaten by 30,000 people.

October 11

In 1987, a record breaking pizza pie was baked. It measured 100 feet and 1 inch in diameter. It was cut into 95,248 slices. **How many slices did each person get?**

How is a pizza like a bowling ball?
How would this large pizza look to a balloonist overhead? An ant on the ground?
Predict which pizza topping is the most liked in your class. Graph the results as a class.
Survey people outside of your class to determine their favorite pizza. Share your results in at least 2 different ways.

October 12

Harvey Gilston, a record bird watcher, was born in 1922.

A bird watcher is called a *"twitcher."* Mr. Gilston, a record breaking *"twitcher,"* reported seeing 6,891 different bird species by April 1991. **If there are 9,016 known species of birds, how many more birds does Harvey need to find?**

List names of birds that have one syllable...two syllables...three syllables.
List different ways birds can be categorized.
Draw a mural of birds in their habitats.

A gigantic yo-yo was built in San Francisco.

A yo-yo is a small toy made of two round pieces of wood that are connected with a peg. A string is attached to the peg and wound around it. By tying the end of the string to your finger, you can do tricks with the yo-yo. Early yo-yos were used for hunting and fighting. In 1929, Donald Duncan watched a demonstration of the yo-yo and decided to make a toy. In 1979, Dr. Tom Huhn built a yo-yo that was 256 pounds. He had to use a 150 foot crane to test the toy. **If a regular yo-yo weighs 4 ounces, how many more ounces is the gigantic yo-yo?**

List different names for a yo-yo. Name other toys or games that use string. How is a yo-yo like jelly? Invent a game to play with a yo-yo. Play it with a friend. Have a trial to decide if the yo-yo is a safe toy.

October 14

The first supersonic airplane flight was flown today.

Supersonic airplanes fly faster than the speed of sound. Captain Chuck Yeager of the United States Air Force flew the first supersonic jet over Edwards Air Force Base in California. The plane was called "Glamorous Glennis" and flew 42,000 feet above the earth. Captain Yeager went 670 miles per hour. **If a car travels at 60 miles per hour, how much faster does the jet travel?**

CHALLENGE If the "Glamourous Glennis" flew 2680 miles, how long would it take to cover the same distance by car at 55 miles per hour?

Name things that fly.
List other names for "Glamorous Glennis" the supersonic airplane.
List things that are super.
How would you feel if you could ride in "Glamorous Glennis?"
What would a jet say to a spaceship?
Make a clay model of a jet or spaceship.

October 15

Let's celebrate World Poetry Day.

Some of the earliest poems were so important they were used as textbooks in school. Write a poem about all the ways to describe the number 15.

What would happen if your math textbook was written as a poem? How is a poem like a supersonic jet? Write a poem about the history of your town or city. Collect favorite poems from other countries and make a class poetry book. Select your favorite poem and illustrate it.

Celebrate Dictionary Day.

Noah Webster was born in 1758. He was a teacher and a lexicographer. He wrote *An American Dictionary of the English Language*. It was published in two volumes. We celebrate this day in his honor. How long are most words? To find out, you will need a newspaper article. Make a tally of the length of words from one letter to ten letters. Which word length occurred most often? Least often? Try your experiment again with another article. Are the word lengths the same or different? Why or why not?

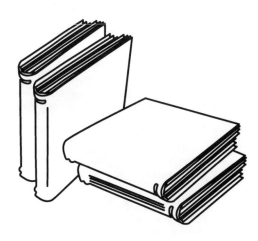

Combine your most favorite and least favorite words.
Make new words from them.
Write a dictionary entry and definition for your words.

October 17
Black Poetry Day is celebrated.

Jupiter Hammon was one of the first black American poets to publish his own verse in 1911. Many of his poems were very religious. His first poem had 88 lines. **Estimate how many pages Hammon's poem was if each page had about 21 lines.**

Write a poem describing a minute. Illustrate your poem. Start a class collection of poems centered around a theme. Using puppets, act out your favorite poem from the collection.

October 18

Alaska was purchased by the U.S. in 1867.

Alaska was purchased for less than two cents per acre from the Russians. In 1959, Alaska became the 49th state. Alaska also ranks about 49th among the states in population. **When will Alaska celebrate its 200th year as a state?**

CHALLENGE Alaska is 586,400 square miles. How much larger or smaller is your state?

Write a letter convincing a friend to live in Alaska.
Alaska has a volcano named Katmai. How many other U.S. states have volcanoes? What are their names? Make a chart of these volcanoes.
Which country has the greatest number of volcanoes? Least number? Collect current disaster stories to share with the class.

October 19   Ed Emberley, author and illustrator, was born in 1931.

Ed Emberley's first book, **The Wing on a Flea: A Book about Shapes,** was published in 1961. Many of his books show children how to draw using only nine easy shapes. Find one of his drawing books. Create a picture using Emberley's nine shapes.

Do fleas really have wings?
Which shape are you most like? Why?
How does a circle make you feel? A rectangle? A triangle?
Using only one shape, add details to complete a picture.

October 20

The Circus opened!

The circus, known as *"The Greatest Show on Earth,"* began in 1873. The show brought many different animals to perform in front of an audience. Did you know that an elephant can weigh 23 times as much as a lion? **If a lion weighs 373 pounds, how much could the elephant weigh?** Round your answer to the nearest ten.

What kind of circus animal are you most like? least like?
What would an elephant in the circus ask an elephant in the wild?
What would a lion in the wild ask a lion in the circus?

October 21

Grete Waitz ran the New York City Marathon.

Grete Waitz was the first woman runner to run a marathon in less than two and one-half hours. She ran a distance of 26 miles and 345 yards in 1979. It took her 2 hours, 27 minutes and 33 seconds. **How much less than 2 1/2 hours did Waitz run the marathon?**

What are some things you run from? What are some things you run to? How is running like reading a book? List books with or about running. Write your own rebus running story. What is a triathlon? Set up a class triathlon. Design a medal for the winner.

Sam Houston was sworn in as President of Texas.

Houston became President of the Republic of Texas in 1836 and then again in 1841. It was Sam Houston who helped Texas become a state in 1845. At least twice in his life he lived with the Cherokee Indians. Houston stood nearly six feet tall. Using pictures from newspapers and magazines, make a mobile or collage of items taller than six feet.

Houston was President twice and lived with the Indians twice. What things have happened twice in your life?
Houston, Texas, was named after Sam Houston. What other U.S. cities were named after famous people?
Choose one city and role play the famous person that city was named after.
Have the class guess who you are and in which state the city is located.
What other states have a city named Houston?

A heavy brain broke a record.

In 1975, the heaviest brain known belonging to a man was recorded this day. It weighed 4 pounds 8 ounces. The brains of humans are getting heavier. The average male brain has increased from 3 pounds to 3 pounds 2 ounces. Women's brains have increased from 2 pounds 11 ounces to 2 pounds 12 ounces. **How much more than the average man's brain did this heaviest brain weigh?**

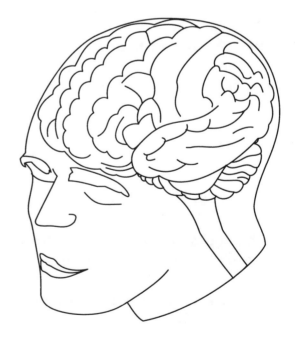

The brain acts as the "mission control center" of the body.
List other kinds of control centers.
Mindmap the topic "brain."
Compare and contrast the brain with a queen bee.
Write a riddle about the brain.

The Pony Express stopped its mail service.

Pony Express riders rode 20 miles in 59 minutes. Mail going west from St. Joseph, Missouri, was delivered in 10 days. Mail going eastbound from Sacramento, California, was delivered in 11 1/2 days. This was the fastest way for mail to travel. The service ran both day and night through a wild and dangerous territory. Only one rider was killed! This service ended in 1861. The charge was $5 for each 1/2 ounce packet. **How much would it cost to send a packet that weighed 2 ounces?**

CHALLENGE How much would it cost to send a packet that weighed one quarter of a pound?

How would you feel if you were a Pony Express Rider? How would you feel about the Pony Express if you were a settler waiting for a letter? an Indian who lived near the route of the riders? Name things that are carried and what or who carries them.

October 25

Two women investigators graduated from the FBI.

Susan Lynn Roley and Joanne E. Pierce graduated from the FBI in 1972. FBI investigators are called special agents. The name, FBI, was given by Congress in 1935. What do the letters FBI stand for? **List abbreviations used in mathematics.**

What do FBI agents do that uses numbers? List other fields or occupations that use abbreviations. What would be a good abbreviation for you? your best friend?

Steven Kellogg was born. October 26

Did you know that Beatrix Potter and W.C. Wyeth wrote and illustrated some of Mr. Kellogg's favorite books? Steven Kellogg, born in 1941, especially loved animal books. **If you borrowed and read three of Kellogg's books from the library on Monday, twice as many on Wednesday, and 1/2 as many on Friday as you did Wednesday, how many would you have read?**

Compare and contrast Kellogg's and Potter's books.
Choose one of Kellogg's or Potter's books in which the characters make a journey or trip. Make a map by drawing the routes for the characters.
Design paper cutouts of the characters. Describe the journey or trip to a friend using your map and cutouts.
What has been your longest journey? your shortest journey?

President Teddy Roosevelt was born.

A cartoon show explaining how Teddy Roosevelt saved a bear cub by refusing to shoot it on a hunting trip made the stuffed teddy bear popular. During Roosevelt's time as President, he established 51 national bird reservations, five national parks, four game refuges, one national game refuge and 18 national monuments. **How many sites were created?**

Do you think Teddy Roosevelt was an animal lover?
Why or why not?

Read five stories with bears. Choose one and make a bear mask to help tell the story.
Invite someone from a national park, monument, or refuge to talk to the class about the importance of preserving these areas.
Write a letter to a newspaper expressing your concerns.

Captain Cook was born.

Cook, one of the greatest explorers, was born in 1728. He sailed around the world twice and explored many islands in the Pacific Ocean. He also sailed the Antarctic waters where he saw many icebergs. The largest iceberg seen in the Antarctic was 200 miles long and 60 miles wide. **What is its length and width in kilometers?**
(There are 1.6093 kilometers in a mile!)

*Some of Cook's ships were named **Adventure, Discovery** and **Endeavor.** Do you think these were good names? Why or why not?*
If you were an explorer like Cook, what would you name your ships? Why?
Compare and contrast an iceberg to a candle.

Artificial rain was created to drench a forest fire.

Cumulus clouds were seeded with dry ice to encourage rain drops to fall. This artificial rainfall helped put out the forest fire. **If Sunny Susie, weatherwoman, predicts a 20% chance of rain tomorrow, what is the chance that it will not rain? If Sunny Susie predicts a 20% chance of rain for the next 10 days, how many days do you expect it to rain?**

CHALLENGE Check the weather in your area for the next two weeks. What is the probability of rain?

Make a mobile of things lighter than a cloud.
Create your own folk tale about the first rain.
List reasons why you like rainy days.
List reasons why you dislike rainy days.

A patent for the ball-point pen was awarded.

John Weymouth of Massachusetts obtained a patent for his ball-point pen in 1888. Many pilots in World War II used ball-point pens because they didn't leak when flying high above the earth. Today, more than 2 1/2 billion ball-point pens are made in the U.S. every year. **Write the numeral for 2 1/2 billion.**

How is a pen like a rubber band?
List other uses for a ball-point pen.
Which was invented first — the pen or pencil?
Which would you rather write with — a pen or pencil?
What 5 questions would a pen from 1888 ask an erasable pen?

1 _____

2 _____

3 _____

4 _____

5 _____

Juliette Low was born today.

Juliette Low organized the Girl Scouts of America on March 12, 1912. During her wedding, a grain of rice fell in one ear. After it was removed, she found that she was deaf in that ear. Some time later, she lost most of her hearing in the other ear. That didn't stop her from founding the Girls Scouts. Four years after founding the Girl Scouts, the first Brownie troops were organized. **What year was it?**

New girls joining the Brownies or Girl Scouts make a promise.
What are some synonyms for promise?

_____

_____

Why are promises made?

_____

_____

_____

Why is a promise like a rainbow?

_____

_____

_____

If you were a promise, how would you feel if you were broken? Write an essay "The Broken Promise."

_____

_____

_____

_____

_____

_____

_____

October Challenge

Write a multi-step word problem using addition and multiplication for the picture below.

_____

_____

_____

_____

_____

KDKA radio station goes on the air.

Radio is an important communication system. A radio program begins in a station where voices and music are changed into electronic signals and sent through the air to radios. In the radio, the signals are changed back into sounds you can hear. In 1920, KDKA station began in Pittsburgh. This first program gave information about the presidential election of Warren Harding. **How many years has it been since KDKA started broadcasting?**

What is a commercial radio station? Name things that have sound. List things that travel through the air. Compare and contrast television and radio. What if there were no television sets? Create a class radio program.

Daniel Boone was born in Pennsylvania in 1734.

Daniel was an early explorer, trapper, and pioneer where he grew up in Pennsylvania. The Indians taught him how to survive in the woods. Later as an adult, he explored Kentucky and helped "blaze" the Wilderness Trail. After the Revolutionary War, he moved to Missouri where he died on September 26, 1820. **How old was he when he died?**

Many pictures of Daniel Boone show him wearing a coonskin hat which became his trademark. List other trademarks and what they stand for.
List survival skills for living in the woods; living in the Antarctic. How would you feel if you were exploring a river for the very first time?
Create a diorama of pioneer days.

John Montague, Earl of Sandwich, was born in 1718.

According to the story, John Montague was the first person to put meat between two slices of bread and make a "sandwich." **If you took six sandwiches to school, gave two to your teacher and half of what was left to your friend, how many would you have left? How many slices of bread did you use?**

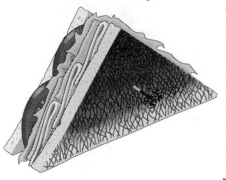

How is a sandwich like a basketball?
List types of sandwiches.
Create a new sandwich and bring it to school for lunch.
Develop a flowchart that shows how to make your sandwich.
Identify criteria of a good sandwich and have a jury of your peers judge your sandwich.

King Tutankhamen's tomb is discovered.

Archaeologists are like detectives who search for clues to discover what life was like long ago. They find clues from uncovering large structures such as houses, tombs, and forts or from small objects like tools, coins, or fragments of broken pottery. The articles that are found are studied and then placed in museums for people to see. On November 4, 1922, the tomb of Egyptian King Tut was discovered. It was one of the most astounding archaeological finds. British archaeologist Howard Carter uncovered almost 2,000 priceless objects including jewelry and gold figures in King Tut's tomb. **If each object was worth $500, estimate how much these objects were worth.**

What is the study of archaeology? What other things do detectives search for? How is a clue like a sandwich? If you were King Tut, what would you put in your tomb? Draw a diagram of King Tut's tomb. Make a mobile of "artifacts" that might be found in an Egyptian tomb.

"Ma", a female tabby, lived to the age of 34.

Cats are mammals that range in size from domestic cats to tigers. All breeds of cats have short, round heads with long whiskers. They also have sharp teeth that they use for grabbing and biting. Cats use their claws and tails for climbing, jumping and balancing. Domestic, or tame, pet cats usually live an average of 11 years. Mrs. Alice St. George Moore, of Devon England, claimed that her pet cat "Ma" lived for 34 years. The cat died on November 5, 1957. **In what year was "Ma" born?**

List different types of cats. What would you say to a saber-toothed tiger? What cats live in families? You are a pet store owner— recommend a breed of cat for a customer who wants a pet cat. Design a poster that shows why a cat makes a good pet.

The inventor of basketball was born in 1861.

In 1891, James Naismith, a physical education teacher, invented the game of basketball. He wanted to invent a game that could be played indoors in the evening. Early basketball hoops were peach baskets and the "basketball" was a soccer ball. It wasn't until 1893 that players used metal hoops for baskets. In 1936, basketball became an Olympic sport and 22 countries entered teams. **How long after basketball was invented did it become part of the Olympic games?**

How is a basketball like a peach? Name things that bounce. Hold a panel discussion on what would happen if you played basketball with eleven players on a side instead of five. How would the game of basketball be different if it were played on the moon? Write a science fiction story about the game.

The first cash register was patented.

In Dayton, Ohio, James J. Ritty invented a device to keep his clerks from stealing his money. Mr. Ritty attempted to sell his invention but no one was interested. He finally sold it for $1000. **Write out the numeral $1000.**

What are the attributes of a cash register?
Compare and contrast an adding machine and a cash register.
How have cash registers changed since Mr. Ritty's invention?
Design a piggy bank that will keep your money safe from anyone trying to "break in."

The Statue of Liberty, a present to the United States from France, is begun.

Today in 1875, Frederic Auguste Bartholdi began working on the Statue of Liberty and finished it on April 22, 1886. It was made of copper, weighed 225 tons and was 151 feet tall. After the statue was finished it was taken apart and sent to the United States. *The Lady,* as the statue was called, was put back together piece by piece and placed on a pedestal on Liberty Island in New York Harbour. The statue stood in the harbour for many years. Eventually the weather started to ruin her. To celebrate the 100th anniversary of the Statue of Liberty, the people of France and the United States raised $265,000,000 to repair her. **If the celebration occurred on July 4th, one hundred years after she was finished, what was the date of the celebration?**

Name things that can be taken apart and put back together.
List other famous statues.
What might the Statue say to people who are coming to the United States? Why is the Statue of Liberty called a lighthouse?
Write a letter to the President telling why the Statue should be refinished.

The great Northeast blackout occurred today.

Electricity is the power that heats and lights our homes. In 1965, an electric power failure caused "lights out" in seven Northeast states and Ontario, Canada. Over 30 million people were left in the dark. The blackout lasted for 13 1/2 hours. **If the lights went out at 5:27 p.m., when did they come back on?**

CHALLENGE If you spend at least 1/4 of each day sleeping, how many total days a year do you spend sleeping?

List other compound words, like blackout, that name a color. Name things that electricity powers. What would happen if the lights went out in your house?
Design a comic strip about what might have happened during the blackout.

If there's a blackout, how can I drive my electric car?

In the 1800's, electric cars, powered by batteries, were popular because they were clean, quiet and easy to operate. As people traveled further and further distances, the batteries were not powerful enough to make the long trips. **If a battery-powered electric car could travel 5 miles in an hour and a gasoline-powered car could travel 5 miles in a half hour, which is faster and by how much?**

Name things that have power.
Compare and contrast a gasoline-powered car with a battery-powered car.
What if cars had not been invented? Create a flip book of a gasoline motor working.

A record-breaking, rare whale shark was caught.

One of the largest sharks in the ocean is the whale shark. It can measure over 50 feet long. The whale shark will not harm people or animals because it eats *plankton*. This shark is found in warm areas of the Indian, Atlantic and Pacific Oceans. A record-breaking whale shark weighed 33,600 pounds and measured 23 feet around the thickest part of its body. **If the smallest shark, the dogfish, is only 2 feet long, how much longer is this whale shark?**

What would a shark in the ocean say to a shark in an aquarium?
Draw a picture of a whale shark and label its parts.
Write a picture story about two different kinds of sharks.
Share it with a younger reader.

A gigantic iceberg was sighted in the Antarctic.

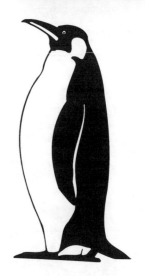

Antarctica remains very much like it was when early explorers discovered the continent. It is covered with giant icebergs. These icebergs are parts of glaciers that have broken off and floated into the sea. Most of the iceberg is under the sea and cannot be seen. In 1959, a huge iceberg was sighted 1,150 miles west of Scott Island in the South Pacific. The berg measured 200 miles long and 60 miles wide. **How many square miles was the iceberg?**

How is an iceberg like a butterfly? Would you like to live in the frozen land of Antarctica? Why? Why not? What question would you ask a penguin if you met one? Make a sculpture of an iceberg.

Robert Louis Stevenson was born in 1850.

Robert Louis Stevenson, born in Edinburgh, Scotland, was one of the most successful and popular writers. Although he had a law degree, his first love was writing. He wrote children's poetry, travel books, novels and short stories. One of his most famous books is *A Child's Garden of Verses*, a book of poetry for children. His first novel was a story he told his stepson about the buried treasure of Captain Kidd. In 1883, *Treasure Island* was published. **How old was he when his first novel was published?**

Name things that are treasures.
Read a poem from A Child's Garden of Verses and illustrate it
or write your own children's poem and illustrate it.
Compare and contrast one of Stevenson's poems with a nursery rhyme.

"Bluey" was a very old dog when it died in 1939.

People have had dogs as pets or work animals for many years. There are more than 100 breeds of dogs today. Most dogs live to be between 8 and 15 years of age. *Bluey* was a sheep and cattle dog that lived on a ranch in Australia. It lived to be 29 years and 5 months old. **If an average dog lives to be 12, how many more years did *Bluey* live?**

Name things that are old.
List different breeds of dogs.
How would the world be different if there were no dogs?

Pandas look like bears, but scientists say that their nearest relatives are raccoons. There are two types of pandas. One is very large and looks like a black and white bear. The other is small and red. Both live in the bamboo forests of China. Pandas have a sixth finger that looks like your thumb. These bears eat a lot of food. They can eat 70 pounds of food in a day. **If two bundles of bamboo weigh 1 pound, how many bundles could a panda eat in one day?**

How are pandas like raccoons?
Why is an animal considered an endangered species?
Are pandas endangered? Why? Why not?
Create a class mural of a panda family in their natural habitat.
Write a skit about your panda family and use the mural as a backdrop.

The first manmade object VENERA 3 was launched to another planet on this date in 1965.

VENERA 3, a Russian spaceprobe, landed on Venus on March 1, 1966. It was spaceprobes from Russia and the United States that gave us some of the very first information about this planet. **Estimate how many months it took VENERA to reach the planet.**

What southern state has a city or town named Venus? _____

List other planets that have cities or towns named after them.

_____

Locate these cities and towns on a map.
Which one is closest to you?

_____

Farthest away?

_____

Choose one you would like to visit and explain why.

_____

_____

_____

The longest journey of the Olympic torch began today in 1987.

The longest journey for the Olympic torch began when it left Greece and finally arrived in Calgary, Canada, on February 13, 1988. The torch traveled 5088 miles by foot and 4419 miles by aircraft and ferry. Another 1712 miles was added to the torch's journey by snowmobile and three miles by dogsled. **How many kilometers did the torch travel?** Round the answer to the nearest thousand.(1 mile = 1.6093 kilometers)

Plan your own metric Olympics measuring sports such as paper plate discus, straw javelin, and broad jump.
Invent other new sports to be measured metrically.
Compete with another class.

Mickey Mouse celebrates his birthday.

Mickey appeared in 1928 in *"Steamboat Willie."* This was the first animated sound cartoon picture. Goofy and Donald Duck were later added to the Mickey Mouse family of animals. A mouse like Mickey may eat up to 63 pieces of cheese in 3 days. **If he eats ten more pieces each day than he did the day before, how many pieces did he eat each day? How many pieces of cheese will he eat on the fourth day?**

Mickey Mouse has many human-like qualities. Who is he most like?
Fill in this blank.
Mickey Mouse is most like _____
because _____

_____

Create a riddle about mice. _____

_____

_____

_____

_____

Answer:

George Rodgers Clark celebrated his birthday.

Clark was a frontiersman and soldier who was born in 1752. As a young man he was a surveyor. His younger brother was best known for exploring the land west of the Mississippi. Do you know his name? What is a surveyor? What are the tools of a surveyor? **If Clark surveyed a rectangular piece of land with a width 1/4 the distance of the length of 336 feet, what is the distance of the width? What is the perimeter of the rectangle? The area?**

Which is your favorite shape?
Describe it to a Martian.
Draw a picture or use cut outs of different sizes of your favorite shape to make a Martian.

Chester Gould, creator of comic strips, was born in 1900.

Gould created the famous comic strip *"Dick Tracy"* on October 4, 1931. He drew the comic about the detective, Dick Tracy, until 1977. Many new inventions like the two-way wrist radio and television and closed circuit TV were used in *"Dick Tracy."* **Measure the perimeter and area of one box of your favorite comic strips found in the newspaper.**

Read some comic strips from the newspaper.
Which comic strips use inventions of the future?
What would Chester Gould say to Charles Schulz?
Make a list of questions Charlie Brown would ask Dick Tracy.
1.

2.

3.

World Hello Day is celebrated!

Over 142 countries celebrate World Hello Day. Say hello to at least 10 people to help celebrate peace. Did you know that the world grows by at least three people every second? **How many more people on the earth do we have to say "hello" to after one minute?**

Look up ways to say "hello" in other languages.
Invent a different way to teach them to your friends.
Write your own book about "Hello."
Make a bookmark that welcomes a reader to a book.

Wiley Post, barnstormer, was born in 1898.

His airplane was named the *Winnie Mae.* Post was also known as a stunt parachutist. Barnstormer Post taught himself to fly. He was the first person to fly solo around the world in 1933. His flight took 7 days 18 hours and 49 minutes. Two years earlier, Post and a friend flew the *Winnie Mae* around the world in a record 8 days 15 hours and 51 minutes. **How much time did Post save in his solo flight?**

What is a barnstormer?
List things that require courage.
What are some synonyms for courage?
What is the difference between encourage and courage?
How is courage like the Sahara Desert?
Draw a picture of a time when you felt like you needed more courage.

November 23

Boris Karloff was born in 1887.

Mr. Karloff was best known for his role in the movies *"Frankenstein"* and *"The Mummy."* **If he made about 50 silent films and 100 sound films during his lifetime, how many movies did he make?**

Complete the following sentence. "I feel scared when . . . .
What other words describe things that are scary?
What does a monster look like?
Using a variety of materials, take turns in small groups creating a new monster.
Each person has two minutes to add to the creation.

November 24   Barbed wire was patented in 1874.

Joseph Glidden of Illinois started to make barbed wire in November of 1873. Barbed wire is made up of two or more steel wires twisted together. Thornlike barbs are placed throughout the wire. **How much wire is needed to fence in a flower garden that is 9 feet long and 3 feet wide?**

CHALLENGE *Measure the area and perimeter of your silhouette.*

Which flower are you most like? Why?
How would our lives be different if we didn't have flowers?
Flowers symbolize spring to some people. What symbolizes spring to you?
Design a flower garden that you could plant in the school yard. Label the flowers and describe their care.

DRIVE
SAFELY

The speed limit was lowered in 1973.

The speed limit was lowered from 70 miles per hour to 55 miles per hour. This saved 2.4 billions of gallons of gasoline per year. In 1987, the speed limit for rural roads increased to 65 mph. **How many more miles per hour could motorists now travel on rural roads?**

CHALLENGE *How many zeros are in a thousand? million? billion?*

Many driving rules are represented by shapes and pictures which are symbols. An R X R placed in a circle warns of railroad tracks.
What symbols are used in mathematics?
Create two symbols to represent something in mathematics.

November 26

Mary Walker, first female U.S. Army surgeon, was born in 1832.

On November 11, 1865, Dr. Mary Walker received the Medal of Honor. She died on June 3, 1916. A special postage stamp was made marking the anniversary of her birth. **How old was Walker when she received the Medal of Honor?**

A surgeon treats patients by operating on them. Using a phone book, list different kinds of doctors.
What is different about what they do? Which kind of doctor would you choose to be? Why?
Draw a Medal of Honor for a teacher.

November 27

"Old Billy" died in 1822.

The oldest recorded age for a horse was 62. "Old Billy" was born in 1760. His skull is preserved and on display in the Bedford Museum, United Kingdom. The height of a horse is measured in hands. Find out how many inches are in a hand. **How tall is a horse that measures 14 hands?**

Name other famous horses and tell why they are famous.
Mindmap the topic of "horses."
Compare and contrast the importance of horses in our past, present and future.
Make a time line of the history of the horse.

The first skywriting appeared.

In 1922, the first skywriting was seen by people who lived in New York City. Predict which letters might have occurred most often in the skywritten message. To check your prediction choose a book. Tally how many times each letter of the alphabet was used in several pages. Which letter is used most often? Least often?

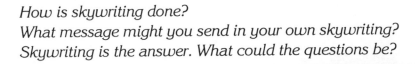

CHALLENGE What percent of our alphabet is consonants?

How is skywriting done?
What message might you send in your own skywriting?
Skywriting is the answer. What could the questions be?

November 29

Charles Thomson, official record keeper, was born in 1729.

It was Mr. Thomson, official record keeper, who let George Washington know that he had been elected President of the United States. Thomson arrived in America from England as an orphan when he was 10 years old. **What year was it?**

List things that are recorded. Categorize your list. Why is it important to keep records? Name people who keep records. Write a riddle or a pun about records.

November 30

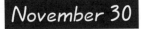

Samuel Clemens was born.

Samuel Clemens, who used the pen name Mark Twain, was born when Halley's Comet was seen in 1835. Twain is the author of **Tom Sawyer.** He died when Halley's Comet was seen again in 1910. **How many years took place between the two sightings of the comet? In what year will we next be able to see Halley's Comet?**

When was this comet first sighted and recorded?
Could Mark Twain have viewed Halley's Comet through a telescope?
How many comets have been sighted during your lifetime?
How would you find out?
Why did Clemens use a pen name?

72 is the Answer

Create 10 equations whose answer is 72.

1

2

3

4

5

6

7

8

9

10

Gingerbread house built for Christmas.

In 1985, a hotel in Arlington, Texas, began to construct a gigantic gingerbread house for Christmas. The house itself was made of 650 gingerbread *"bricks"* and 195 *"shingles."* The inside of the house was lined with 100 pounds of dark chocolate. The chefs used over 1,000 pounds of icing to decorate the outside. When the house was finished, it stood 17 feet high and almost 3 yards wide. The doors to the house were large enough to allow young children to enter. **How many days did it take to build the house if every day 20 people worked on it for 8 hours (160 hours) and it took 1,120 hours to complete?**

CHALLENGE Work with a partner. Cut out four 2" x 2" squares. Arrange all four squares to form a structure with the perimeter of 24 inches.

How is a gingerbread house like a horse? Compare and contrast a gingerbread house with an Easter egg. Design your own gingerbread house and list what each part would be made of. Estimate the cost of actually making your house.

December 2

Emperor Bokassa of Central Africa wears an $85,000 pair of shoes.

The Emperor had a pair of pearl-studded shoes made for his self-coronation in December of 1977. Another pair of expensive evening shoes was produced by Reinhard Seufert of West Germany. His shoes were made of gold leather and decorated with an emerald brooch on one shoe. The shoes, with rhinestone screws, are valued at about $50,000. Mr. Seufert rents them for stage and television productions. **How much more do the Emperor's shoes cost than Mr. Seufert's shoes?**

CHALLENGE How do shoe sizes and body weight compare? Collect the shoe size and weight of at least ten children and adults. Record your information on a graph. What are the results?

What things did early people use for shoes? List kinds of shoes made today and where you would wear them. What if shoes had not been invented? Design a pair of shoes for a special occasion. Write a paragraph about the shoes, who wears them, and what the occasion is.

December 3   "Benedictine Jr Schwarzwald Hof weighed in at 310 pounds.

Benedictine was born in 1982 and weighed 310 pounds on December 3, 1984. He measured 39 inches tall at his shoulder. St. Bernards are the heaviest breed of dog and have an average weight of 170 pounds. The Saint Bernard is one of the working dogs and came from Switzerland. These dogs were trained to help find stranded skiers in the Alps. The smallest dog is the Yorkshire terrier which weighs less than a small jar of jam. **If Benedictine weighed 2 pounds at birth, how much more did he gain until he was weighed on December 3, 1984?**

Take a survey to find out the class's favorite dog. Make a chart of their favorites.
Create a crossword puzzle using different "dog" words.

The bloodhound is another "dog" helper.

Bloodhounds belong to the class of working dogs. They have a terrific sense of smell. This ability is what helps them find missing children and even track down criminals. Working dogs, such as Collies, old English sheep dogs and Shetland sheep dogs, help farmers herd sheep and lambs. Siberian huskies, Alaskan Malamutes and Samoyeds pull sleds. Other dogs such as Doberman pinschers and German shepherds serve as watch dogs. Besides having a good sense of smell, dogs can hear very well. They hear some sounds 250 yards away. **If a person can only hear the same sound from 25 yards away, how many times further away can a dog detect sounds?**

Names things that track and what it is they track. List things that working dogs could do. How is a clue like a bloodhound? Draw a picture of a working dog doing his job.

Happy Birthday, Walt Disney!

Walter E. Disney was born in Chicago. Better known as Walt Disney, he is famous for producing animated cartoons featuring characters such as Mickey and Minnie Mouse, Pluto, Donald Duck, Goofy and others. He made full-length cartoon movies such as *"Snow White," "Cinderella," "Pinocchio"* and *"Bambi."* Disney also made regular movie features including *"Treasure Island," "Robin Hood"* and *"20,000 Leagues Under the Sea."* In 1932, he produced the first cartoon in technicolor. Later, he filmed wildlife in their natural surroundings and combined real life people with cartoons. **If Walt Disney was born in 1901 and died in 1966, how many years did he live?**

List cartoon characters. Name things that move. How is Pinocchio like a tree? Create a dialog between you and Pinocchio. With a partner share it with the class.

Lionel trains are a popular toy.

Joshua Lionel Cowen made his first toy train of wood when he was seven years old. When he grew up, he made a model railroad flatcar that ran with a motor. He sold it and 30 feet of railroad track for $6.00. Soon he began making many different types of railroad cars. **If he sold 100 cars with 30 feet of track at the same price as his first sale, how much would he make?**

CHALLENGE One half of Mr. Smith's 30 students is boys. Two-thirds of the class like model trains. How many like model trains?

Name different types of railroad cars and what they carry.
How is a railroad train like a Greyhound bus?
Would you like to be an engineer on a railroad train from New York to Kansas City?

The phonograph is demonstrated for the first time in 1877.

Thomas Alva Edison invented the phonograph along with over 1,100 inventions in his lifetime. When he was 7, Thomas Edison was sent to a formal school. He stayed only three months. His mother spent many hours teaching him at home and making learning fun. With money from one of his first inventions, he opened a workshop in Newark, New Jersey. He later moved to Menlo Park, where he worked on his favorite invention, the phonograph. Edison invented a better electric light. His was cheaper, brighter and longer lasting than others. Edison died October 18, 1931 at the age of 84. **In what year was he born?**

Name things that are records.
List 5 ideas Mrs. Edison might have taught Thomas.
Draw an invention that browns bread efficiently.

Eli Whitney is born in 1765.

Eli Whitney was born in Massachusetts, the son of a farmer. He always liked to make things as he was growing up. After he graduated from Yale, he moved to Georgia to study law. To pay for his room and board, he fixed things around the house. His landlady asked him to make a machine that would clean cotton. His cotton gin amazed people because it cleaned 50 pounds of cotton a day. It had taken 50 people one day to clean 50 pounds by hand. **How many times faster than one person was the cotton gin?**

List things that are made from cotton.
Name things that are white.
What if the cotton gin had not been invented?
Design a product that could be mass produced by your class.
Produce the item and market it.

Clarence Birdseye, originator of frozen food, was born in 1886.

On a trip to Labrador, he observed that frozen fish were good to eat after they had been thawed and cooked. Mr. Birdseye developed a method to quick-freeze foods. In 1925, he marketed quick-frozen fish. He also invented a way to dehydrate food. He sold his frozen food idea to General Foods Corporation. In his lifetime, he acquired over 300 patents. **If you went to the grocery store and bought frozen peas for $.89, frozen French fries for $1.89 and a package of frozen fish for $2.39, how much would you owe?**

List things that are frozen. How is a frozen fish like a light bulb? What if there were no refrigerators? Write a news story about Mr. Birdseye's newly discovered frozen food. Design an attractive package for his frozen fish.

December 10

The first Nobel Peace Prize is awarded in 1901.

Alfred Nobel, a Swedish chemist, found a chemically safe way to produce a new explosive he called dynamite. He wanted it to be used for peaceful uses. When he died, Nobel left instructions that his $9,000,000,000 estate be used to give awards to the best work in the fields of chemistry, physics, medicine, literature, economics and outstanding efforts in working toward international peace. The first person to receive the Nobel Peace Prize in 1901 was Jean Henri Dunant. He founded the Red Cross and helped organize the Geneva Convention. **For how many years have the Nobel Prizes been awarded?**

List things that "blow up." Name things that are peaceful.
How is dynamite like an automobile?
If you could ask Mr. Nobel a question, what would it be?
Create a new award for school citizenship. Select criteria for awarding the prize. Design the "prize."
Award it monthly to a deserving classmate.

Annie Jump Cannon, American astronomer, was born.

December 11

Astronomy is the oldest science on earth. It is the study of the stars and heavenly bodies in the sky. Early astronomers studied the sky by looking up at the stars. Later telescopes were developed to help scientists view the sky. Space probes, like Viking I and II, have also brought information about our neighbors in space. Annie Jump Cannon discovered 5 stars. **Out of 125 stars, what is the percent of stars that Annie discovered?**

What stars did she discover? What star can you see during the day?
How does a star look to a comet? an airplane? a meteorite?
How is a star like a butterfly? Draw a picture of what the sky looks like at night. Label any stars or constellations.

The first transatlantic wireless signal is sent in 1901.

Guglielmo Marconi was born in Bologna, Italy. He read and studied a lot as a young boy. Mr. Marconi was an Italian inventor who investigated sending messages through the air without using wire. This idea led to today's radio communication system. In 1901, he built a sending station in England and sailed to Newfoundland to set up a receiving station. His first message traveled across the Atlantic Ocean. It was the letter *"S."* Marconi worked on this device so that it could send more wireless signals over greater distances. **If a message took 3 minutes to go 80 miles, how long would a message sent 240 miles take?**

CHALLENGE What percentage of an hour did the message take?

Name things that have wires. List things that are wireless. How is a radio like an orange?
What would you have sent as your first message across the ocean?

Get ready for a "lifesaving" candy record.

Thomas Syta of California kept a lifesaver candy in his mouth for 7 hours and 10 minutes with the hole still intact. Lifesavers are one of the most popular candies and have sold over 34,000,000,000 rolls since being invented in 1913. **If a package of lifesavers has 20 candies in it and sells for $.60, how much does each lifesaver cost?**

Name things that are lifesavers. What other things can you do with lifesaver candies besides eat them? Experiment with dissolving lifesaver candies. What method works best? Draw a chart with your results. Design and market a new candy.

George Washington died at Mount Vernon in 1799.

George Washington was the first President of the United States. He is sometimes called the *"Father of our Country."* There are many memorials to Washington. Streets, towns, cities, a state, bridges, schools and many other things have been named in honor of him. One spectacular monument that he shares with three other Presidents is called Mount Rushmore. The faces of the four Presidents are carved on a granite cliff in the Black Hills of South Dakota. Who are the other Presidents carved on the mountain? **If work on the monument began in 1927 and was finished in 1941, how long did it take to build?**

List famous monuments.
What do you think the Presidents might be saying to each other as they look down from the mountain? What would you like to ask one of the Presidents in the monument? Compare and contrast the Mount Rushmore monument with the Statue of Liberty.
Design a monument to someone that you admire.

Today is Bill of Rights Day.

In 1947, Franklin D. Roosevelt set aside December 15th as a day to observe the original signing of the Bill of Rights. The Bill of Rights is the first 10 amendments to the original United States Constitution. They were written to make Americans more aware of their rights and responsibilities as United States citizens. **How many years has it been since Roosevelt announced the Bill of Rights Day?**

Name things that have bills.
List things that have rights.
Draw up a Bill of Rights for your classroom, outlining the students rights and responsibilities in class.
Create a chart with the Bill of Rights on it and post it in the room.

Our Classroom's Bill of Rights

1. _____

_____

_____

_____

2. _____

_____

_____

_____

3. _____

_____

_____

_____

4. _____

_____

_____

The first museum for children opens.

The Brooklyn Children's Museum in New York opened in 1899. The museum houses more than 50,000 items in what is called a "teaching" collection. There is a greenhouse, a steam engine and a gristmill for children to explore. There are now over 200 museums built especially for children. **If a museum has 56 exhibits and 3 new exhibits are added each year for the next four years, what will be the total number of exhibits?**

You are to name 3 new exhibits at the Brooklyn Children's Museum. What will they be?
Design a poster that would encourage children to visit one of your exhibits.
Is it a good idea to have museums for children?
Why or why not?

December 17

The first airplane flies at Kitty Hawk.

Orville Wright was the first to fly in their biplane, *"Flyer,"* after winning the toss of the coin with his brother, Wilbur. **How many ways could the one tossed coin have landed? Two tossed coins? Three tossed coins?**

CHALLENGE If three coins are tossed, what is the probability of 3 heads? 2 heads?

List other famous brothers and reasons why they are famous.
Describe the "perfect" brother or sister.
What other things are tossed besides coins?

December 18

The first daguerreotype of the moon was taken in 1839.

John Draper took the image of the moon which was one inch in diameter. He presented his photograph on March 23, 1840. The moon circles the earth once every month. **How many years would it take the moon to circle the earth 60 times?**

CHALLENGE On which of the following celestial bodies—the earth, moon, sun, or Mars—would you weigh the most? the least? Why?

The days of the week are named after the sun, moon, and five planets. What day of the week is the moon named after?
What is your favorite day of the week? Why?
Illustrate how you feel about each day of the week.

Richard Leakey, anthropologist, was born.

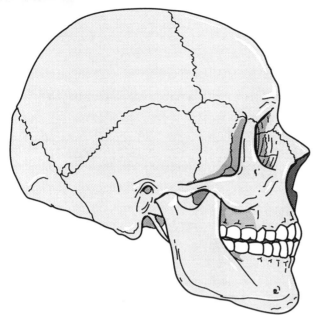

Both of Richard's parents were also famous anthropologists. Leakey's parents found a human skull almost 1.75 million years old. Richard found another human skull almost 3 million years old. **Write the numerals for 1.75 million and 3 million.**

Compare and contrast an invention with a discovery.
Would you rather invent something or discover something?
How is an invention like an iceberg?

Sacagawea died.

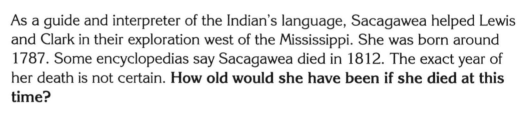

As a guide and interpreter of the Indian's language, Sacagawea helped Lewis and Clark in their exploration west of the Mississippi. She was born around 1787. Some encyclopedias say Sacagawea died in 1812. The exact year of her death is not certain. **How old would she have been if she died at this time?**

Her Shoshone name meant "Bird Woman." Why is this a good name?
Think of other names that would fit Sacagawea.
What if Lewis never met Clark?
If we all had numbers instead of names, what would Sacagawea's number be? What would your number be?

Celebrate National Flashlight Day!

Do you know why we would celebrate National Flashlight Day? Could it be the longest night of the year? This day could have been celebrated since 1898 when the first dry cell battery flashlight was made in New York. **If five classes of 22 students took a flashlight tour of the art museum, how many flashlights were needed?**

List times when flashlights are needed. How is a flashlight like a life preserver? What other uses can you think of for a flashlight? Using a flashlight, create an image of something on the wall. Ask a friend to guess what it is.

Colo the gorilla was born.

Colo was the first gorilla born in captivity in Columbus, Ohio, in 1956. Colo weighed 3 1/4 pounds. **How many ounces did Colo weigh?**

Make a chart showing the weight of animals less than 2 kilograms. Place them in order from lightest to heaviest.
The answer is gorilla. What could the questions be?
Write a riddle, pun, or joke about gorillas.

How is a gorilla like a garbage can?

The Metric Conversion Act was passed.

The metric system finally became a legal system of measurement in the U.S. in 1975. **Estimate in centimeters the height of the girl. Estimate her weight in kilograms. How many meters to you think she can run in 15 minutes?**

Design an ad encouraging other children to use the metric system.
Invent a game that teaches younger children about the metric system. Play it with some younger children.

Kit Carson was born today.

Christopher Carson, born in 1809, was a frontiersman, trapper and guide. He was famous for being able to track anything on two or four feet. **If Kit was tracking an animal with footprints 3 inches apart and he followed 16 prints, how far did the animal run?**

Create a new footprint. Describe who made it and the reason why it was made.
Kit was described as brave, honest, gentle and wise. Describe yourself using each letter in your name to begin words that tell about you.

Clara Barton, nurse, celebrates her birthday.

Clara Barton, born in 1821, became the first president of the American Red Cross. She was 60 years old. **In what year did she become president?**

What kinds of things does the Red Cross do?
Write to them for more information.
What other associations or groups help people?
Choose one and compare and contrast it to the Red Cross.
Think of another name for the Red Cross.
What will you be doing when you are 60 years old?
How would you feel about the Red Cross if you were a doctor?
a survivor of a hurricane?
a leader of a poor country?

Second Day of Christmas is observed.

This Second Day of Christmas is celebrated in many countries. **How many more days from this date until you celebrate your birthday?**

Many towns and cities in the United States have names of Christmas holiday words like North Pole, Arkansas; Santa Claus, Indiana; Rudolph, Ohio.
Use an atlas to find other city names that remind you of the
holiday season. List the city and state below.

_____   _____

_____   _____

_____   _____

_____   _____

_____

Rewrite "The Night Before Christmas" and change it to "The Night After Christmas."
Illustrate your new version.

The world's tallest identical twins were born.

James and Michael Lanier were born in 1969. At the age of 14, they each measured 7 feet 1 inch. They now stand 7 feet 4 inches. Identical twin sisters, Heide and Heather Burge, measured 6 feet 4 3/4 inches tall. **Estimate how much taller the Lanier twins were than the Burge twins.**

What is the difference between identical twins and fraternal twins?
Make a collage of things that come in pairs.
What do we call things that come in threes? fours? fives?
How would it feel to have a twin?
Would you rather have an identical twin or a fraternal twin?

Pairs or Twins?

Chewing gum was patented.

William Finley Semple of Mount Vernon, Ohio, obtained a patent for gum in 1869. **If William tested 10 pieces of gum on Monday, half as many on Tuesday, 8 pieces on Wednesday, three less on Thursday than on Monday and 16 pieces on Friday, how many pieces of gum did he chew?**

Bubble gum allows chewers to blow large bubbles. "Chiclets" were invented to give chewers a candy coated piece of gum.
Invent a new kind of gum. What would you name it?
Survey at least ten classmates. What is their most favorite gum? Least favorite?

The inventor who used bonded rubber to keep dry was born today.

Charles MacIntosh was born in 1776. His invention led to the use of bonded rubber to make raincoats. But, on July 4, 1956, many children did not have time to put on raincoats during the greatest one minute rainfall in Maryland. Over one inch of rain fell in just one minute. **How many inches of rain would have fallen in 2 1/2 minutes?**

Describe how a rainy day makes you feel.
A snowy day?
A sunny day?
Write a poem about one of these days.

_____

_____

_____

How would your life be different if rubber raincoats had not been invented?

December 30

The Monitor Sank.

The *Monitor* was a famous battleship built of iron and used in the Civil War. Steam was used to power this ship. The *Monitor* sank while being towed during a storm at sea on December 31, 1862. Underwater archaeologists have located the ship off the coast of North Carolina. Its anchor was found 121 years later. **In what year was the anchor found?**

List things that give you a sinking feeling.

_____

How is an anchor like an organ?

_____

List other uses for an anchor.

_____

_____

Illustrate one time when you had a sinking feeling.
Design a new way to keep a ship from moving away from the dock.

December 31

Celebrate "Make Up Your Mind" Day.

Celebrate this day by making a decision and sticking with it! **If it took 50 days to make up your mind about something, how many weeks would have passed?**

What are the chances that a decision may be right? wrong?

_____

_____

Why are some decisions hard to make?

_____

_____

_____

List decisions that are hard for you to make.

_____

_____

_____

_____

How is making a decision like making a bed?

_____

_____

_____

Which is easier to make—a decision or a phone call? Why?

_____

_____

_____

_____

December Challenge

This Bear has a story about his problem! It's a MATH story problem!
Make up his story problem using three math operations.

_____

_____

_____

_____

_____

_____

_____

Answer

The first self-propelled fire engine was tested today.

Cincinnati, Ohio, was the site of the first self-propelled fire engine. It was powered by a steam engine. Before this, men pumped the water needed to put out the fire. A pumper fire truck carries water tanks, a hose and a pump to pump the water. A ladder truck carries ladders to reach fires that are in tall buildings. Today fire engines run on gasoline and pump great quantities of water.
If a ladder truck and two pumper trucks left for a fire and then minutes later, twice as many ladder trucks and a pumper truck left, how many trucks would arrive at the fire?

What do they call fires that need many trucks and firefighters? How is a fire engine like a balloon? List other ways to put out a fire. List things that are opposite of fire. Design a lesson to teach young children about fire safety rules.

Today is Good Luck Day.

Today is considered a day when good luck will come to everyone. You just have to look for it. Some people think that finding a four-leaf clover is good luck. Most clover plants have only three leaves. Clover plants are found growing wild in fields, lawns and along the highways. Bees use some clover for making honey, and farmers use clover to enrich their soil. **How many "leaves" are there in a group of 16 four-leaf clovers?**

List other things that mean "good luck." If you could have one good luck wish, what would it be? Why? What numbers are considered lucky? Create a rebus story about a lucky day .

Today is National "Sip-a-Drink-Through-a-Straw" Day.

The wax drinking straw was patented in the United States in 1888. Today we celebrate this invention. A box of straws contains 24 straws.
How many boxes of straws do you need to serve sodas with straws to everyone in your class?

List all the things you can do with a straw.
What if the straw had not been invented?
Compare and contrast a straw and a strawberry.
Using only three straws at a time, construct as many different shapes as you can.

Today is Jacob Grimm's birthday.

Jacob Grimm, a librarian, was born on January 4, 1785, in Germany. Jacob's main interest was in writing a dictionary of the Germanic language. But he and his brother, Wilhelm, are most famous for their collections of fairy tales. The two brothers collected and wrote down the stories that people told them. Their first book was published in 1812 and called *Household Stories*. **How old was Jacob when he published his first book?**

CHALLENGE If the Grimm Brothers wrote 2 pages a day and their average story was 18 pages, how many days would it take them to write 20 stories?

Compare and contrast two of Grimm's fairy tales.
How are numbers important in fairy tales?
Make a chart that shows 3 different fairy tales and any important numbers in them.
What if the Grimm brothers came to your house for dinner? What questions might you ask them?
What stories might they write down about your family?

Today is National Bird day.

In 1905, the National Audubon Society was founded to encourage people to preserve wildlife. The Society conducts an annual census that counts the number of bird species that live in each part of the country. The largest bird is an ostrich and the smallest is the hummingbird. Most birds do not fly higher than 3000 feet above the ground; however, geese have been seen flying over the Andes Mountains at a height of 29,000 feet. **How much higher can these geese fly than most birds?**

Which birds do not fly at all?
Describe different bird habitats.
What do birds eat?
Compare and contrast the habits of hummingbirds and ostriches.
Would you rather be a hummingbird or an ostrich?
Research the types of birds and their nesting habits in your area and present your findings in a chart.

Carl Sandburg was born in 1878 in Galesburg, Illinois.

He was a poet, biographer and historian. His most famous biography was about Abraham Lincoln. It took him 20 years of writing and researching to finish **Abraham Lincoln**. One of his poems was called "Fog." He described fog as moving like a cat. In 1951, he received a second Pulitzer Prize in poetry for his book **Complete Poems. How old was he when he received the second Pulitzer Prize?**

List ways to describe fog. How is fog like a cat? What names might Mr. Sandburg call a cat?
Draw a picture of a foggy day and what you would be doing. Write a diamante poem about fog and sun.

January 7

Fannie Farmer published her first cookbook in 1896.

Fannie Merritt Farmer was an American cooking expert. She directed her own school for cooking in Boston. She taught her students about diet, cooking and kitchen management. Her famous cookbook, called the **Boston Cooking School Cook Book** was in its 21st edition when she died in 1915. Today there is a book entitled **Original Boston Cooking School Cook Book**—a facsimile of the original book published in 1896. **How many years after her death is this book still being sold?**

What does facsimile mean? Bring to class and share with your classmates your family's favorite recipes. Make a class cookbook.
Create titles for your cookbook and choose the best ones.

January 8

"George", a very tall giraffe, arrived at England's Chester Zoo in 1959.

When George was 9 years old, his horns almost touched the roof of his giraffe house. The zoo keepers recorded his height at near 20 feet which made him one of the tallest giraffes in the world. He died in the zoo on July 22, 1969. **How many years did George live in the zoo?**

Name as many zoo animals as you can.
Which animal has the longest life span?
Which is your favorite zoo animal? Why?
Choose two different zoo animals. Put them together to create a new animal.
Draw a picture of what it would look like and write a description about it.

The first successful balloon flight in the US was in 1793.

Frenchman Jean Pierre Blanchard had flown more than 40 balloon flights in Europe. He was the first to fly over the English Channel. He thought that Americans would pay to see a balloon flight. He charged people $5.00 to watch. With his small black dog aboard he launched his balloon from inside a prison in Philadelphia. Most people stayed outside the prison walls and watched for free. **If 100 people had paid to see him, how much money would he have made?**

How would you feel if you were flying in a balloon over the countryside?
over the ocean?
over an active volcano?

How would you look in the balloon to a bird flying by?
to a plane flying over you? to your friend on the ground?
Create an advertisement to encourage people to attend this event.

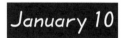

January 10

The London Underground Railway opened its first section.

The earliest underground railway was in London. The first part of this subway system was opened today in 1863. Underground railroads are called *subways*. There are subways in New York City, London, Paris and Moscow. **How many years has the London Underground operated?**

How is a subway ride like a balloon flight?
How are they different?
What could you see on the subway that you could not see in a balloon?
Would you rather ride on a subway or a balloon?
Design a ticket for the grand opening ride on the London subway.

Today we celebrate Banana Boat Day.

The banana shaped dish that holds a banana split sundae was invented today. The banana plant looks like a tree and grows in warm countries. Bananas grow in bunches and are picked while they are still green. **If each member of your class wanted a 3-scoop banana split sundae, how many scoops of ice cream would you need to fix the sundaes? How many bananas would you need?**

What other things could you put in a banana boat?
Create a new ice cream and banana dessert.
Draw a flowchart to show the steps in making your dessert.
Create a catchy name and write a jingle to advertise the dessert.

The first American museum was opened.

In 1773, the first American museum was opened in Charleston, South Carolina. Museums collect and preserve original objects. These objects might be paintings, automobiles, or prehistoric animals. The word museum comes from Greek and means *"a place to study."* **If a museum held 2000 things and 50 went in each room, how many rooms would you need?**

Name things that you might find in a museum.
What things from your classroom could be donated to a museum in order to describe what your class is like to someone in the future?
Create a class collection of your favorite hobbies. Label the collection.
Write an invitation and include a special ticket inviting a younger class to visit your "museum."

Today is Steven Foster Memorial Day.

Steven Foster was a famous American songwriter. Even though he did not take music lessons, he was very musically talented. When he was 6 years old, he taught himself to play the clarinet. In 1846, he wrote the song *"Oh Susannah"* about the forty-niners in the California gold rush. He wrote over 200 songs but sold them for very little money. Even though his songs were popular, he died a poor man. **How much money would he have made if he had sold each song for $3.00?**

Pretend you are a California forty-niner. Tell about your day.
Name things that are gold.
How is a song like a gold nugget?
Act out one of Stephen Foster's songs.

The Cerro Aconcagua volcano was climbed.

Volcanoes are openings in the earth where burning gas and rocks escape. Sometimes these materials trickle out of the opening and other times they explode. When they explode, scientists call it a volcanic eruption. The Cerro Aconcagua volcano, nicknamed the *"Stone Sentinel,"* was the highest extinct volcano to be climbed in 1897. It is in the Andes Mountains next to Argentina, and it is 22,834 feet high. **How much taller is the volcano than you are?**

CHALLENGE How many people of your height (round off height to the nearest foot) lying end to end would it take to stretch 22,834 feet?

How are you like a volcano? Not like a volcano?
What other names might you call the "Stone Sentinel?"
Name things that erupt.
Create a web about volcanoes.

Dr. Martin Luther King was born in 1929.

Dr. Martin Luther King, a famous American Civil Rights leader, was born in Atlanta, Georgia. He wanted to have all people live in peace with each other. He won the Nobel Peace Prize when he was 35 for his leadership in the struggle for equal rights. **In what year did Dr. Martin Luther King win the Nobel Peace Prize?**

Name other men and women who have worked for equal rights.

_____

_____

_____

_____

Compare and contrast their deeds with Dr. Martin Luther King's.

List peace symbols.   *Design a peace symbol. Draw it here and explain its meaning.*

_____ _____

_____ _____

_____ _____

_____ _____

Create a crossword puzzle using synonyms for peace.

Celebrate National Nothing Day.

National Nothing Day was first observed in 1973. Howard Coffin, an American newspaperman, felt that we needed one national day when we could just sit and relax without doing any special celebrating or honoring. **How many years has this day been observed?**

Holidays are special days that honor important events or a famous person. What is your favorite holiday? If you could create a new holiday, what would it be? Make a poster about National Nothing Day or your new holiday. Create a joke or riddle about nothing.

A famous puppeteer and ventriloquist celebrates her birthday.

Shari Lewis, puppeteer and ventriloquist, was born in 1934. Lewis's favorite puppet, *Lamb Chop*, has been with her many years. Other animals Shari brought to life include *Charley Horse*, a nag, and *Wing Ding*, a bird. Shari's birth date, January 17, can be written as the sum of the consecutive numbers 8 and 9. For example 8 +9 =17. Consecutive numbers are numbers that go in order like 1, 2, 3 or 10, 11.

CHALLENGE Which numbers between 1 and 20 can't be written as the sum of two consecutive numbers?

What is a ventriloquist?
How do you become a ventriloquist?
Invent a new puppet for Shari.
What name would you give the puppet?
Create a script for the puppet.

The author of the Winnie the Pooh books is born.

A.A. (Alan Alexander) Milne was born in 1882 and died in 1956. He is best known for his books **Winnie the Pooh** and **The House at Pooh Corner.** Winnie the Pooh has adventures that take place in a forest area called the Hundred Acre Wood. **In what century was he born? In what century did he die?**

Compare and contrast Winnie the Pooh with a real bear.
What questions would a bear in the circus ask a bear in the wild?
Which would you rather be — the circus bear or the bear in the wild?

January 19

James Watt, inventor, was born today.

James Watt was a Scottish engineer born in 1736. One of his inventions was the modern steam engine. The power unit, called the watt, is named in his honor. A light bulb shows how much power it needs by the number of watts. **If a room needed 150 watts of light and you had three lamps that you must use, how many different combinations of the following bulbs could light the room - 25 watts, 40 watts, 50 watts, 75 watts, 100 watts, and 125 watts?**

Which is brighter—a light bulb or a rainbow?
If you could spend one day with an inventor, who would it be? Why?
How would life be different if Watt had never been born?

January 20

The second person to step on the moon celebrates his birthday.

Edwin Aldrin, astronaut, was born in New Jersey in 1930. "Buzz" Aldrin was the second person to set foot on the moon on July 20, 1969. It took 3 hours to put on life support equipment. He stepped out of the Eagle 19 minutes after Neil Armstrong. **If Armstrong first placed his foot on the moon at 10:56 P.M., what time was it when Aldrin followed?**

Footprints on the moon went only 1/8 inch deep. Why?

How would you have felt if you were the first person to step on the moon?

Which planet in the solar system would you most like to visit?

Draw a picture of the first and last day of your visit to the planet.

January 21 — Celebrate National Hugging Day.

Official huggers should spend the day hugging friends and family. **If you gave 10 hugs to friends in the morning, 8 hugs in the afternoon, and 9 hugs in the evening, how many hugs did you average for each part of the day?**

Which would you rather hug—an ostrich or a giraffe?
How is a hug like a good book?
Create a commercial which explains the importance of hugging friends and family.

January 22

"Answer Your Cat's Question" is celebrated today.

Today is the day when owners take a serious look at their cat to try and figure out what question it is asking. Many cats are very attached to their owners. Take the case of *Sugar*, who refused to travel in the family car and was left behind when his family moved from California to Oklahoma. Although a neighbor adopted *Sugar*, he disappeared. Thirteen months and 1500 miles later, *Sugar* showed up in Oklahoma! **If there are approximately 30 days to a month, how many miles did *Sugar* travel each day to reach his family?**

What questions would you ask Sugar about his journey? List new names you could give Sugar that would provide clues about his adventures.
How would you feel if you were Sugar trying to find your family?
Describe in a story or picture a day's adventure for Sugar.

January 23

Elizabeth Blackwell received her doctor's degree in 1849.

Elizabeth and her sister Emily, also a doctor, opened their own hospital in New York City. It was called the New York Infirmary. Most of the people helped in the hospital were poor. A medical school for women was later added to the infirmary. In 1949, the Elizabeth Blackwell Medal was established in her honor. **If this medal has been awarded each year since 1949, how many women have received it?**

Which would you rather be—a doctor or a veterinarian? Why?
List other health occupations. Interview someone in one of the occupations.
Make a diorama of what that person does during the day.

Eskimo Pies are patented!

Christian Nelson of Iowa received a patent for Eskimo Pies in 1922. Eskimo Pies are vanilla ice cream bars dipped in chocolate. About 7 quarts of milk are required to make 4 quarts of ice cream. Did you know that Americans eat more ice cream than people from any other country?

If the average American family eats 16 quarts of ice cream in a year, how many quarts of milk are needed?

What ices, other than ice cream, can you think of?
Which causes you to shiver more—ice cream or a scary movie?
Write a riddle that has something to do with ice cream. Share it with a friend.
Create a new recipe in which ice cream is one of the ingredients.

The first monument created by a woman is shown.

Vinnie Ream was paid $10,000 to complete a marble, life-sized statue of Abraham Lincoln. It was unveiled in the Rotunda of the U.S. Capitol in 1871. **If the material for creating the monument cost $5660, how much did Vinnie Ream make in profit?**

What other monuments have been created by a woman?
Select a National Monument.
Make a poster or brochure encouraging more children to visit the monument you selected.
Make a short presentation to the class about the monument you chose to represent.
How would you feel about that monument if you were an eagle?
a park ranger?
a camper?

Celebrate National Popcorn Day!

Today we honor popcorn. The average kid eats about 33 quarts of popcorn. When corn kernels are heated to a temperature of about 400° they burst with a pop into fluffy flakes. This is because there is a tiny bit of water in the middle of the kernel. When it gets hot, it releases steam. The kernel expands 30 to 40 times its original size. **How big would a kernel .2 centimeters be if it expanded 30 times its normal size?**

Compare and contrast the word "pop" with the word "expand." What other things pop? Expand? Create a picture using popcorn. Give it a title.

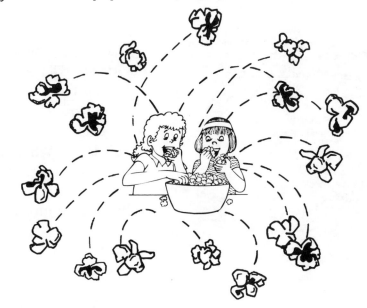

January 27

281,581 out of 320,236 dominoes were toppled in 1984.

Most of the original dominoes set up by Klaus Friedrich in West Germany had fallen in less than thirteen minutes. He spent 31 days setting up the dominoes. Each day he spent 10 hours getting the dominoes set up. **How many total hours did he spend setting up the dominoes?**

CHALLENGE What percentage of dominoes toppled?

What other things can be toppled? List reasons why things topple. Create a new game using dominoes. Teach it to a friend.

National Kazoo Day is celebrated.

A kazoo is a musical instrument which changes your voice into a buzzing sound when you hum into the mouth hole. The sound is similar to one made with a comb and paper. **What body parts can be used to make a pattern of sound?**

CHALLENGE What musical notes are written in terms of a fraction?

Name other uses for a kazoo. What would it be like if people buzzed instead of talked? Invent a "buzz" alphabet. What things buzz? Create a pattern song using these sounds. Teach it to a friend.

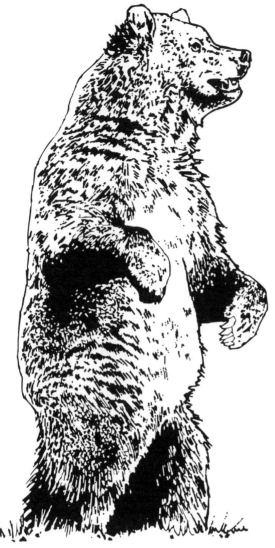

Author Bill Peet was born in Indiana.

Bill Peet's first book was published in 1959. He has written and illustrated over 25 books, many of them animal stories. His favorite animal book was **Wahbi: The Biography of a Grizzly**. Bill Peet also worked for Walt Disney for 27 years. He was the writer and illustrator for "101 Dalmatians" and other films. Open one of Peet's books so that two pages facing each other are 6 and 7. The sum of these two pages equals 13. **What two pages would you need to open so that the sum of the pages equals 49?**

What other movie titles contain numbers?

_____

_____

_____

Fill in the blanks: Number _____ is like a _____ because

_____

_____

_____

Choose your favorite number. Illustrate this number in a picture.

The first two-way moving sidewalk was opened.

This sidewalk began its service at Love Field Air Terminal in Dallas, Texas. It was 1435 feet long. **How many yards did it save people from walking?**

CHALLENGE Find the average length of your stride. Do this by marking a starting spot on the floor with tape. Take ten steps from the piece of tape. Mark it with another piece of tape. Measure the distance between your starting and stopping steps. To find the average length of your stride divide this distance by 10. Why would it be useful to know the length of your stride?

Which is longer—a moving sidewalk or an embarrassing moment?
What is your most embarrassing moment?
Draw a flowchart of the events that led up to this moment. Could it have been avoided?
What is the busiest airport in the United States? in the world?
Mind map the word "airport."

Jackie Robinson, baseball player, was born today.

Jackie Robinson played with the Brooklyn Dodgers from 1947-1956. In 1949, he was voted the National League's Most Valuable Player. In 1962, he was the first black person elected to the Baseball Hall of Fame. A special postage stamp was issued in his honor in 1982. Jackie Robinson died on October 24, 1972.

If a worker can turn out a new bat for Jackie and his team every eight minutes, how long would it take to make 50 new bats?

Challenge Thirty-six baseball cards are dealt out to a group of friends with none left over. When a new friend joins the group and the thirty-six cards are dealt, one is left over. How many friends now have baseball cards?

How are new stamps selected?
In the box below design a postage stamp in honor of your best friend.
Write a letter to the Post Office explaining why they should use the stamp you designed.

Dear Mr. President,

January Challenges

The answer is 10. How many DIFFERENT math problems can you create?

Pretend you are an inch worm. List all the things you COULDN'T measure.

List all the words you can think of that are part of the study of mathematics.

You are a math detective. How many uses of math in your life can you detect? List and explain.

Compare and contrast addition, subtraction, multiplication, and division.

You have just been elected to the Math Hall of Fame. Why? Describe your math talent.

Collect several jigsaw puzzles. Calculate the average time it takes you and your friends to put together a 100, 500, and 1000 piece puzzle. Graph the results.

Would you rather solve ONE hard math problem or 30 easy ones? why?

The "Guinness Book of World Records" asks you to create the world's longest, most unusual word math problem. Try it!

From a Good Source   **Active Questioning** by Nancy Johnson

© 1995 Pieces of Learning

Today is "Be an Encourager Day"

This is a special day for you to encourage and help your family and friends. It is also the 32nd day of the year. **How many days remain in this year?**

Write all the ways that you can be an encourager.
List synonyms or words that mean the same as "encourage."
Name one thing that you wish someone would encourage you to do.

Today is called Groundhog Day, and we celebrate the custom of looking for the ground hog's shadow.

Legend says that if the ground hog, or marmot, comes out of his hole and sees his shadow, there will be six more weeks of winter. **If the groundhog sees his shadow today, what will be the last day of winter?**

Why would people make up this legend?
Find a newspaper article about "Groundhog's Day" and bring it to class to share.
What animals hibernate during the winter?
What animals migrate to another area during the winter?
What are some of your favorite things to do in the winter?
Write a legend about this day.

Elizabeth Blackwell, the first woman physician, was born on this date in 1821.

Elizabeth Blackwell established a hospital in New York City with an all female staff. In 1981, a U.S. postage stamp was designed in her honor. **If a first class stamp costs $ .32, how much money would it cost you to send valentines to six special friends?**

What kinds of jobs are there in a hospital?
What kinds of jobs are there in a post office?
Invite someone who works in a post office or a hospital to speak to your class.
Compare and contrast work that you would do in a post office to work you might do in a hospital.
Would you rather work in a post office or a hospital?

Charles Lindbergh, American pilot, was born in 1902.

Charles Lindbergh was the first man to fly from New York to Paris without stopping. His nicknames were *"Lucky Lindy"* and *"Lone Eagle."* He flew alone in his airplane which was named the *"Spirit of St. Louis."* It took him about 33 hours. Today the Concorde, a supersonic jet, can fly from New York to Paris in about 4 hours. **How much longer did it take Lindbergh to fly?**

Make a chart of all your classmates' nicknames. How do you get nicknames?
Are some nicknames better than others? Why? Survey some people who have flown in an airplane.
Construct a graph showing how many times they have flown.

Today is called Weatherman's Day in remembrance of the birthday of one of America's first weathermen.

John Jeffries, the first weatherman, was born in 1744. He kept a lot of information about the weather and wrote down what he observed. Would you be surprised to know that he was also a doctor? The highest temperature ever recorded in the United States was 134° F in Death Valley, California, on July 10, 1913. **What is today's highest temperature? What is the difference between today's high and Death Valley on July 10, 1913?**

CHALLENGE If the low temperature was 45° last Wednesday, and it was 5° lower each day, what will the low temperature be next Wednesday?

Name things that are high. Name things that are low. Name things that are both high and low. Collect newspaper weather graphs for one week. Pick a town and chart the high and low temperature for each day. Display your information in a graph.

Babe Ruth, the first great home run hitter, was born in 1895.

Babe Ruth hit 714 homes runs during his baseball career. Hank Aaron broke Ruth's home run record in 1974 when he hit his 715th home run. He finished his career with 755 home runs.
How many total home runs did these two baseball players hit?

What other records did Babe Ruth hold?
What pitcher holds the record for the most strike-outs?
Who invented the game of baseball?
Using a ball and bat, invent a new game.

Laura Ingalls Wilder, author of the "Little House" books, was born today.

Laura Ingalls Wilder, born in 1867, wrote about her own life in the "Little House" series. In 1954, Laura received the American Library Association's award as an outstanding author. Beginning in 1960, the ALA has presented an award every five years to an outstanding author or illustrator. It is named the "Laura Ingalls Wilder Award" in her honor. **In what years has this award been given? How many times has it been given?**

Read one of the "Little House" books.
Compare and contrast your life today with Laura's.
Which character are you most like— Laura or Mary? Explain why.
Make a chart comparing and contrasting which character you are most like.

Jules Verne Day

Jules Verne wrote two very popular books entitled ***Around the World in 80 Days*** and ***Twenty Thousand Leagues Under the Sea.*** **If the world is 24,000 miles around, how many miles would you have to travel in one day to make it around the world in 80 days?**

What other books can you list that have numbers in their titles?

_____

_____

_____

Read one of these books and explain why the title was appropriate.

Draw a picture showing what it might look like 20,000 leagues under the sea.
How deep is that in feet? meters?

_____   _____

The oldest orangutan on record died at the age of 59.

The orangutan is a large ape that lives in Sumatra and Borneo. The oldest orangutan on record died in 1977. The male stands from 3 to 5 feet tall and weighs 150 to 200 pounds. **Since females are only about half that size estimate how much she might weigh.**

What is the natural habitat of orangutans?
What do they eat?
Write a story about an orangutan family telling about a day in their life.

Alanson Crane's fire extinguisher was patented today in 1863.

A fire extinguisher is a metal container which holds a chemical or water that can be used to spray on a fire. **If paper burns at 363°F and wood burns at 469° F what is the difference in their kindling temperature (how hot you must heat something to make it burn)?**

Fire can be a friend or an enemy. List all the ways that fire is our friend. List ways that fire is dangerous. Write safety rules about fire. Review the safety rules your school has about fire drills and fire escape routes. Draw a map of escape routes for your home.

Today the United States celebrates National Invention Day.

Thomas Alva Edison, a famous American inventor, was born in Milan, Ohio, on this date in 1847. He had 1,093 patents in his lifetime including the electric light bulb and the phonograph. Have you ever heard of an invention called a laser potato peeler? **If a laser potato peeler can peel 3 potatoes per minute, how many potatoes could you peel in two hours?**

Name other inventions that help you everyday.

Pick two inventions and combine them to make a new invention.

Describe what your new invention can do and who would buy it.

Judy Blume's birthday is today.

Judy Blume, popular children's author, was born in 1938. She wrote *Tales of a 4th Grade Nothing, Super Fudge,* and many other books. In *Freckle Juice*, the big green monster was forced to drink 2 quarts of juice 3 times a day. **How many quarts would he have to drink in 1 week? How much is this in gallons?**

Read two or more Judy Blume books. Compare and contrast the main characters in two of the books. Which of these characters is most like you? Create a recipe for a new drink for you and your classmates.

February 13

License plate #8 was sold for $5 million in Hong Kong.

Some people think that the number 8 is a very dull number. This is probably because it follows the number 7 which is considered to be very lucky. The prefix for eight is "oct." For example, an octopus has 8 legs. **If you saw one dozen octopuses, how many legs would they have all together?**

List other words which begin with the prefix "oct" and tell what they mean. Discuss why someone would pay $5 million for a license plate. Design a license plate for your car. Make the numbers mean something to you. Describe the plate and what your design means.

February 14

Oregon became the 33rd state in 1859, and Arizona became
the 48th state in 1912.

Arizona is known as the *Grand Canyon State*. Oregon is known as the *Beaver State* and also the *Pacific Wonderland* because of all the beautiful natural wonders found there. One of these wonders is a very tall mountain called Mount Hood which is 11, 239 feet high. Crater Lake is also in Oregon. It is 1,932 feet deep. **Mt. Hood is how much higher than Crater Lake is deep?**

CHALLENGE If you could climb Mt. Hood at the rate of 1 mile per day, estimate how many days it would take to climb to the top.

What is your state known for? Design a travel brochure listing famous places in your state. In your brochure encourage tourists to visit your state for a vacation.

Today is Cyrus McCormick's birthday.

Cyrus McCormick, born in 1809, invented a horse drawn reaping machine. Before this time, farmers had to use scythes and could only harvest 2 or 3 acres a day. With the new machine, farmers could harvest more than 10 acres of grain. McCormick demonstrated his machine in a Virginia wheat field. **If the field was 120 feet by 100 feet, what was the area of the field of wheat?**

What crops are grown by farmers in your state?
List all the kinds of things you think might be grown on a farm.
List all the animals that might be raised on a farm.
Interview a farmer to see what crops he grows or what animals he raises.
Share the information with your classmates.

The longest traffic jam ever reported occurred in 1980.

The longest traffic jam from Lyons to Paris was approximately 109 miles long. **If the average car measures 18 feet, how many cars were in the traffic jam?** Collect enough old bits of string or yarn to measure 109 feet. Roll the string or yarn into a ball. Measure the circumference.

What other things would measure this distance?
What causes things to jam?
Invent a game for children to pass away the time while caught in a traffic jam.

Geronimo was a famous Apache warrior.

Geronimo was a warrior of the Chiricahua Apache Indians. He died in 1909. Some people say his Apache name was Goyaale which means *"the smart one."* In 1877, the United States government put all Apaches on a reservation; but Geronimo escaped. He led raids on American settlers. He was caught several times and escaped again and again. He finally surrendered in 1885. He was moved to Fort Sill in 1894 and lived there until his death at the age of 70. **In what year was he born?**

What questions would you have asked Geronimo if you could have visited him on the reservation?
List words found in "reservation."
Locate Indian reservations on a map.

Elm Farm Ollie was the first cow ever to fly in an airplane in 1930.

Elm Farm Ollie was milked while flying in an airplane. The milk was put into paper containers and parachuted over St. Louis, Missouri. This feat was accomplished at the St. Louis International Air Exposition. Today people in Missouri celebrate this event by eating a large amount of cheese and ice cream. **If Elm Farm Ollie gave 1 quart of milk every day for a week, how many pints of milk was this?**

CHALLENGE How many ice cream cones could your class eat if each person ate 3 cones a week for 1 year?

What other events might have occurred at the Air Exposition in St. Louis?
What is your most favorite dessert?
Least favorite?
Write a joke, riddle or pun about ice cream.

Nicolaus Copernicus, a Polish astronomer, was born.

An astronomer is someone who studies the planets. Nicolaus Copernicus was born in 1473. He developed the idea (called a theory) that our earth moves around the sun. Before him, astronomers thought the earth stood still and other heavenly bodies moved around it. Copernicus died on May 24, 1543. **How old was he when he died?**

List the planets in our solar system in alphabetical order; in order from the sun.
Draw a chart with the planets correctly placed next to each other.
Invent a new planet.
How is it like our earth? How is it different? What would force us to live there?

A team of men began building a large snowman.

In 1988, a team of men from Alaska, with Myron L. Ace as their leader, began to build *Super Frosty* a giant snowman. They finished *Frosty* on March 5, 1988. He was over 64 inches tall! **How many days did it take them to finish the snowman?**

What other names could you think of for a huge snowman like Frosty?
Write a poem about Frosty, the giant snowman.
Using scraps of paper, create a picture of Frosty in his snowy winter world.

Lucy Hobbs was the first woman to graduate from dental school.

Lucy Hobbs graduated in 1866 from a dental school in Cincinnati, Ohio. A dentist is a doctor who specializes in the teeth and mouth. By the time you are 2 1/2 you have 20 teeth — 10 in the upper jaw and 10 in the lower jaw. A complete set of 32 teeth may develop when you are older. **How many teeth are in each jaw when you are grown up?**

The first tooth brush was made of a small bone. Holes were made in the bone and bunches of bristles were glued into the holes.
Take a survey of the brands of toothbrushes and/or toothpastes used by your classmates or friends. Share your results in a chart or graph.
Create a new way to clean your teeth. Name things that have teeth.

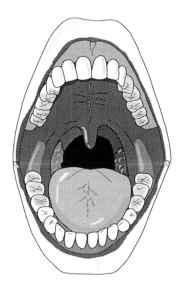

Britain's Boy Scout Founder's Birthday

Today Lord Baden-Powell, the founder of the Boy Scouts of Britain, was born in 1857. Many younger boys also wanted to be scouts; so in 1916 Lord Baden-Powell wrote the Wolf's Cub Handbook which was based upon the story *Jungle Book*. **How old was Lord Baden-Powell when he wrote this book?**

Survey your classmates to find out how many are either Cub Scouts, Boy Scouts, Brownies or Girl Scouts. Make a pictograph showing your information.
What other clubs or organizations do you belong to?
List reasons why boys and girls join clubs.

Johann Gutenberg was born in 1395.

Until Johann Gutenberg invented a way to print from move-able type, all books were written by hand. Gutenberg printed *The Bible* in 1455. It is now one of the most valuable books in the world. This famous book has 1282 pages. **On what page would you be if you had read one-half of the book?**

List advantages of having books. List disadvantages of not having books.
What other ways can people communicate without print-ing letters or words?
What is your favorite book and why?
Design a bookmark for your favorite book.

Wilhelm Grimm, collector of fairy tales, was born in 1786 in Germany.

Wilhelm Grimm and his brother Jacob became known for their many fairy tales. The brothers wrote down the tales just as they heard them from the German people. Three volumes of tales were written. Wilhelm died at the age of 73. **In what year did he die?**

Visit with a relative or an adult who is willing to tell you a story.
Draw a picture that tells the same story without using words.
Using only your picture, tell the story to a friend.
Describe how a good story makes you feel.

Today, in 1956, the heaviest hen's egg ever reported was laid by a white leghorn.

The heaviest hen's egg weighed 16 ounces. This fantastic egg had both a double yolk and a double shell. **If a recipe for a large birthday cake, serving everyone in your class, called for 8 eggs weighing 4 ounces each, how many 16 ounce leghorn eggs could you use instead?**

Make a list of things that come in two's or are doubles.
Is two better than one? Why or why not?
Which number do like better — one or two? Why?

William F. Cody was born in 1846.

William F. Cody, also known as *Buffalo Bill*, was a famous American frontier scout. At the age of 14, he was carrying mail for the Pony Express. When he was older, he hunted bison or buffalo. It has been said that Cody killed 69 buffalo in one day. **How many days would it take him to kill 207 buffalo?**

The buffalo is almost extinct. What other animals are extinct?

_____

_____

_____

Write a poem convincing your classmates to save an animal that is endangered.

Terry Tessman built an enormous bike.

The bike was 72.96 feet long and weighed 154 pounds. The bike was ridden a distance of 807.08 feet on February 27, 1988. **If you left your house riding your bicycle to the library at 3:10 p.m. and didn't return home until 45 minutes later, what time would it be?** Who was the first person(s) to invent the bicycle? Where was it invented?

CHALLENGE If it takes 20 minutes to ride your bicycle to the library and 20 minutes to ride back home again, how many round trips could you make in an hour and 20 minutes?

How is a bicycle like a balloon?
Which would you rather ride—a bicycle, elephant or wave?

The largest seal was killed today in 1913.

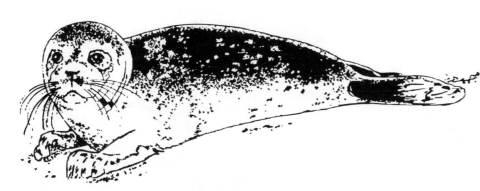

The largest seal ever recorded was found near the Antarctic islands. It weighed over 4 tons and was 21 feet 4 inches long. **If three of these seals lined up one at a time, how long does the ice need to be so that no seal falls into the water? In feet? In inches?**

What words can you use to describe things that are cold?

_____ _____ _____ _____

_____ _____ _____ _____

List animals that live in cold areas. . .   *. . . hot areas.*

_____ _____ _____ _____

_____ _____ _____ _____

Make a collage of your favorite cold and hot things.

People born on this date celebrate their real birthday once every four years.

A year is called a common year when it is not a leap year. A leap year occurs once every four years. It is called leap year because February 29th was not always recognized and this date was "leaped over" in many records. Is this a common or a leap year? **How many leap years will we have before the year 2010?**

Name things that leap and what it is that they leap over.
List words that rhyme with leap.
How would you feel if you had a birthday on February 29th every four years?

February Challenge

Each of the vehicles is able to go a different range of speed. Make up four word problems about speed involving each one of the vehicles.

Today is National Pig Day.

Today is a day to recognize that pigs are one of the most intelligent animals. Besides being intelligent, pigs are cleaner than most other farm animals. Pigs are often called hogs. Pigs or hogs do like to eat. They especially like corn and can gain as much as 1 1/2 pounds every day. On March 5, 1985, "Bud," a Texas pig, was sold for $56,000. He was a very expensive pig. **If a pig usually cost $56.00, how many regular pigs could you buy for the cost of "Bud?"**

Write a conversation between the wolf and one of the three pigs in which the wolf tries to convince the pig why he should blow the pig's house down. Compare and contrast the pig and the wolf. How would you feel if you were one of the three little pigs when the wolf came to visit?

March 2

Dr. Seuss, children's author, was born in 1904.

Dr. Theodor Seuss Geisel was born in Springfield, Massachusetts. One of his first jobs was drawing cartoons. In his books, he has created colorful cartoon-like characters like Horton, Bartholomew, Thidwick the Moose and the Cat in the Hat. Two of his most popular books are *Green Eggs and Ham* and *The Cat in the Hat*. **How old was Dr. Seuss when he died in 1992?**

Name things that are green. How would eating breakfast be different if eggs were really green? What if pigs were green? Read several Dr. Seuss books and compare and contrast the characters in each book. Make a collage of your favorite cartoon characters. Create a cartoon bubble for each character and write how they feel about green eggs.

Alexander Graham Bell was born in 1847 in Scotland.

His family moved to Canada, and he became a teacher of the deaf. He also invented communication devices. Although he is most famous for inventing the telephone, he told his family that he would rather be remembered as a teacher of the deaf. He became a citizen of the United States in 1882. **How old was he when he became a U.S. citizen?**

What other machines help us communicate? How is a telephone like a piano? What other ways can we communicate without using machines?
Create a fact file about machines that help handicapped people. Design a machine to help handicapped people. Explain its function.

The first U.S. National Park, Yellowstone, was established in 1872.

The National Park System has three different types of areas. They are natural, historical and recreational. Yellowstone National Park is the oldest park in the United States. This park has lakes, evergreen forests, meadows, canyons and waterfalls. It is also a wildlife preserve where you can see bears, elk and buffalo. The famous *"Old Faithful,"* a geyser that spouts hot water from the ground about every hour, is found in Yellowstone National Park. **If you were to visit Yellowstone for about 6 hours, approximately how many times would *Old Faithful* spout water while you were there?**

In what three states is Yellowstone National Park?
Research National Parks in your state. How are they like Yellowstone National Park? How are they different?
What would you say right before you erupted if you were Old Faithful?
How do you think the animals feel about geysers?
Create a collage of animals that live in Yellowstone or a park near you.

William Steinway, U.S. piano maker, was born in 1836.

Mr. Steinway was a German cabinetmaker who made the first *"overstrung scale"* piano in the 1830's. This method of stretching bass strings over the other strings greatly improved the piano's sound. Steinway started his own piano company in New York after he immigrated to the United States. Steinway & Sons were famous for making high quality pianos. A Steinway grand piano made in 1888 was sold in New York City for a record $390,000 on March 26, 1980. The piano's buyer did not even play the piano. **How old was the piano when it was sold for this high price?**

How is a piano like a geyser?
How many students in your class play the piano?
Make a chart of the students and musical instruments they play.
Combine two musical instruments and create a new instrument. List its attributes.

Louisa May Alcott, children's author, died in 1888.

Louisa grew up in a very poor family. She left home at 16 to earn money to help support the family. She worked in a hospital during the Civil War. Her first successful novel was **Hospital Sketches** about her experiences. Louisa wrote her most famous book, **Little Women**, in 1868 about her own family. Later she wrote many sequels about the March family. She wrote over ten books before she died. **If she started publishing books in 1854, approximately how many books a year did she write?**

List other stories about families. Compare and contrast a character in **Little Women** *with a character in Winnie the Pooh. How is a book like a window? Write a letter to Louisa May Alcott and tell her about your family.*

March 7

"Mr. Watson, come here I want you."

These were the first words heard over Alexander Graham Bell's telephone. He had been experimenting with his invention with his helper, Thomas A. Watson. For a long time, nothing but noises came over the wire. Suddenly, Graham spilled acid from a battery and called out for help. Watson heard him on the other end of the telephone! Both men forgot all about the mess in their excitement. Mr. Bell received a patent for the telephone on March 7, 1876. **If Mr. Bell was born in 1849, how old was he when the telephone was patented?**

What if there were no telephones? How is a telephone like an airplane? What important words would you have said if you had been Mr. Bell? Write a skit about this important invention and include a scene where the inventors invent something new. Act it out.

March 8

In 1968, Tommy Moore got a hole-in-one on the fourth hole at Woodbrier Golf Course, Martinsville, West Virginia.

A Canadian golfer, C. A. Thompson, shot a score of 103 when he was 103 years old. Tommy Moore was the youngest golfer ever to get a hole-in-one. He was 6 years and 36 days old. His shot went 145 yards. **How much older was Mr. Thompson when he shot 103 than was Tommy when he got his hole-in-one?**

CHALLENGE If Tommy took 4 shots on every even hole and 5 shots on every odd hole, what would he score on 9 holes of golf?

How is a golf course like a ping-pong table? Most golf balls are white; name other things that are white. List things with holes. How would you feel if you were the youngest player to get a hole-in-one? the oldest player? Design a miniature golf course. Develop a brochure to encourage people to visit your course.

March 9 — The "Battle of the Ironclads" occurred in 1862.

Before the Civil War most ships were made of wood. In the famous battle of the *Monitor* and the *Merrimac*, the battleships were made of or covered with iron. The *Merrimac* was originally made of wood and then covered with iron. This gave the ship an armored shield which protected it from shells. The ships fired at each other for approximately four hours but no one was the winner. **If the Civil War began in 1861 and ended in 1865, did this battle occur nearer the beginning, middle or end of the war?**

Name things that are iron. Name things that are wood. Which is stronger—iron or wood? Why? List famous battles. How do you feel when you have had a fight with a friend? your parents? Compare and contrast the Battle of the Ironclads with a joust. Draw a detailed illustration of this famous battle.

March 10

The first United States paper money was printed in 1862.

Early people used shells and beads for money. Today most countries use paper as currency. The United States uses the dollar, but other countries use other units of money. England uses the pound sterling and Japan uses the yen. At first the United States printed only five, ten and twenty dollars bills. **If you get $2.00 a week for an allowance, how many weeks do you need to save to buy something for $20.00?**

How many other paper denominations does the U. S. Treasury print today? What other things besides paper have been used for money? What might we use for money in the future? Collect foreign bills and coins and create a display.

March 11 — Johnny Appleseed died in 1847.

John Chapman, known as Johnny Appleseed, was a pioneer in America and a seed planter. He gave apple seeds and sprouts to his neighbors along the Ohio River. People would give him money, food or other objects to pay for the seeds. They nicknamed him Johnny Appleseed. Apples are a very important food. One apple tree was so good that a nursery owner, in the state of Washington, paid $51,000 for a Starkspur golden delicious apple tree. **If Johnny was born September 26, 1774, how old was he when he died?**

List different types of apples. How is an apple like a quarter? Describe the life of an apple tree in your back yard. Write a fairy tale about what happened to one of Johnny Appleseed's seeds.

Several years after the Boy Scouts were formed, the Girl Scouts of America began. It was founded in Georgia in 1912 by Juliette Low. The Girl Scouts organization is like the British Girl Guides which started three years before in England. Girls between ages 6 and 17 can become scouts. The group encourages members to be honest, helpful, and cheerful. **In what year did the British Girl Guides begin?**

How is a Girl Scout like an astronaut?
What makes a good Girl Scout?
Design a new Girl Scout badge.
Describe how the badge should be earned.

March 13

A very small, cold planet was discovered in 1930.

Pluto is very far away from the earth and sun. It is about 3,675,000,000 miles from the sun. It takes the planet about 248 years to travel around the sun. The earth travels around the sun in one year. Scientists think that the temperature on Pluto is about -184° C. In 1978 another small, cold planet called Charon was found. **How many years after Pluto was discovered did scientists find Charon?**

Name things that are very cold.
How is a planet like a plant?
Describe what life might be like on Pluto.
Create a poem about life on a cold planet.

March 14

A record breaking mural was unveiled today.

A mural is a work of art that is directly painted or drawn on a wall. In 1984, a mural, which was painted on a 30 story condominium, was unveiled in Miami, Florida. The painting had 44 colors and was 300,00 square feet. Some murals stretch a long way. In December of 1984, a mural which was 990 feet long was painted by 6 artists in Birmingham, England. **If each artist painted an equal amount, how many feet did each one paint?**

List as many different colors as you can. What is your favorite color? Why? What if there were no color in the world? How does the color chartreuse make you feel? Draw a class mural about a topic you are studying.

March 15

Today is Buzzard Day.

Buzzards are birds with short tails and heavy bodies. They fly very high and gracefully. They are called birds of prey because they eat small animals like mice and squirrels. This bird has good eyesight and can see its prey from a long way away. The adult turkey buzzard is about 30" tall. His cousin, the black vulture, is about 6 " shorter. The buzzard is part of the vulture family and is sometimes called a turkey buzzard. It is found living from South America all the way to Canada. Each year a group of buzzards return to Hinckley, Ohio, on March 15th. **How tall is the black vulture?**

Name other birds of prey. How is a buzzard like a light bulb? Would you rather be a turkey buzzard or a mouse? Why? Interview a spectator at Hinckley, Ohio, on March 15. Write a news article about the buzzards returning to Hinckley.

March 16

Long-billed curlews arrive.

The curlew, a bird like the sandpiper, has by now arrived at the Umatilla National Wildlife Refuge. This bird has long, skinny legs. The long-billed curlew gets its name from its bill, which can grow to almost 8 inches long. **If 465 of the 500 curlews have bills at least 8 inches long how many curlews have smaller bills?**

CHALLENGE *What percentage of the curlews have bills 8 inches or longer?*

Suppose you were a bird. Which one would you be and why? Design a bulletin board about extinct birds. On a map, locate the areas or regions where these birds once lived.

March 17

Camp Fire Boys and Girls were founded today in 1910.

Each Camp Fire girl or boy is believed special and has lots of talents to share with others. The flame is the Camp Fire's special symbol. It stands for the warmth of the home and the wonder of the great outdoors. **If five friends joined the Camp Fire program and each asked two other friends to join, how many new members would have joined?**

List reasons why we use symbols. Name symbols used in astronomy; in transportation; in medicine. What other fields or occupations use symbols? Design a symbol which would best describe you. Invent a game that teaches about different symbols and what they represent.

11,651 pairs of shoes were shined in 1989.

One of the earliest shoes worn was called a *crackowe*. Some of their toes were so long that they had to be attached to the knee so that the wearer would not trip. Many shoes today are designed by computers. **If it took four teenagers 8 hours to finish shining 11,651 pairs of shoes , estimate how many pairs of shoes each teenager shined. How many single shoes would this be?**

Which would you rather be—a tennis shoe, a ballet slipper, or a crackowe? How is a shoe like a fence? Write a song that would help pass the time if you were shining 11,651 pairs of shoes.

March 19

"Daylight Savings Time" was passed in 1918.

Daylight Savings Time allows us one extra hour of sunlight in the evenings. Daylight Savings Time begins from the first Sunday in April and lasts until the last Sunday in October. As a result, we are able to conserve or save electricity in the evenings. **How many days this year do we get an extra hour of sunlight?**

CHALLENGE *What percentage of the year does not include Daylight Savings Time?*

How is a minute like your favorite dessert? How would you describe a minute to a Martian? Design a flash card game that teaches about time. Play it with a friend.

Celebrate the start of spring!

This day (or the day after) is also known as the Vernal Equinox. The Vernal Equinox takes place when the sun crosses the equator. Spring begins in the Northern Hemisphere of the earth and fall begins in the Southern Hemisphere. Daylight length is just about the same everywhere on the earth — 12 hours and 8 minutes. **How many minutes are in the longest day? Seconds?**

What does the word "equinox" mean? On what day(s) do we celebrate the start of autumn? What is that equinox called? Which is longer—the day or the night? Choose a country you would like to visit. Describe what a child in that country might be doing during 12 hours of daylight.

Today is Fragrance Day.

This is the day to be aware of the different fragrances around you. The nose can recognize as many as 1000 different smells. As much as 10,000 quarts of air flow through the average adult's nose in a single day. **How many gallons of air is this?**

Our sense of smell is related to our sense of taste. Some experts say we smell foods more than we taste them. Describe the smell of an apple, an onion, and a chocolate chip cookie. How are they alike? Different? Compare and contrast the smell, taste and touch of these foods. Share your results in a chart.

The actor, who played the captain of the "Enterprise," was born.

William Shatner, born in 1931, was most famous for his role as Captain James T. Kirk of the U.S.S Enterprise in the original series *"Star Trek."* The program began in 1967 and produced 78 color shows each 50 minutes long. Since the original Star Trek, a cartoon series, film movies and a new television series have been produced. List ten television shows you think your friends like to watch. Ask them to rank their favorite shows from your list. Make a graph showing the results. Which show is the most popular? Least popular?

How would you feel becoming a member of Captain Kirk's crew? Where would you go? What would you see? Draw a picture of your adventures.

A near miss collision with an asteroid occurred.

The asteroid passed within 500,000 miles of the earth. It was a near miss collision in 1989. If the asteroid had collided with our planet, it would have created a crater the size of Washington, D.C. The average distance from the moon to the earth is 238,860 miles. **If the asteroid passed on the other side of the moon, how far away from the moon did the asteroid pass?**

What would have happened if the asteroid did collide with the earth? How would you have felt as a scientist studying the asteroid? an endangered animal that lived in the area? a circus owner bringing her or his show to town? Write a letter to the editor of a newspaper from one of these points of view.

The longest sand castle on record was made.

The longest sand castle on record was over 5 miles in length. It was completed in England in 1988. The record for the tallest sand castle was 17 feet 6 inches high. **How many kilometers is the 5 mile castle?**

Name other words containing the word "sand." List other uses for sand. Which would you rather live in—a castle, teepee or igloo? Why? If you could design your own house tell what it would look like and make a sketch or model.

George Washington planted pecan trees.

Pecans were planted by George Washington at Mount Vernon in 1775. Pecans are also known as America's own nut and were grown by Indians many years ago. The pecan tree is a type of hickory tree. Pecan trees begin to bear a few nuts after 5 or 6 years. Most of the pecans we eat come from trees 20 years old. **If Washington began planting two trees every hour beginning at 8 a.m. and finishing by 6 p.m., how many trees did he plant?**

If Johnny Appleseed met George Washington what would he say? Role play the conversation. List different kinds of nuts. Categorize your list. What is your favorite kind of nut?

A monument to Popeye was uncovered.

The statue of Popeye in Crystal City, Texas, stood 6 feet tall and was made of concrete. When Popeye first appeared, the sale of spinach rose 33%. Spinach gave Popeye his strength. A mathematical error which stated spinach had 10 times more iron than it did led to this belief. Although spinach contains an average amount of iron, we now know spinach has folic acid which does give strength to people eating it. **What other ways can you write 33%?**

CHALLENGE What other vitamins and minerals are contained in spinach? Draw a pie chart showing how much of these vitamins and minerals are present.

Spinach, liver, and broccoli are usually named as the top three least favorite foods of kids. Survey your class or other classes. Do their tastes agree with most kids? Would you rather be Popeye or Superman?

The composer of the lyrics to "Happy Birthday to You" was born.

Dr. Patty Hill Smith was born in 1868. She and her sister, Mildred, wrote the words to *"Good Morning to You"* which became *"Happy Birthday to You."* Mildred was born in 1859. **How old were the sisters when they published the words to this birthday song in 1934?**

Create a new song using the same melody as this birthday song by changing the words to make it different. Share it with the class. How is a birthday like a balloon?

A washing machine was invented.

Nathaniel Briggs was granted a patent for a washing machine on this date in 1905. The first automatic washing machine was introduced in 1937. It could clean up to 9 pounds of wash. **Estimate how many pairs of jeans weigh 9 pounds. Check your estimate at home.**

Before the invention of the washing machine, how were clothes cleaned?
What might a washing machine say to a clothes dryer?
Create a script for a radio advertisement introducing this washing machine in 1937.
How have washing machines changed since then?
Would you rather wash clothes using a modern washing machine or clean the carpet with a broom? Why?
Interview your grandmother. Ask how "cleaning machines" were different when she was growing up.

A yo-yo weighing 820 pounds was launched.

This toy was first introduced in the U.S. from the Philippines in 1929. Yo-yo, in the Philippine language of Tagalog, means *"come come."* In 1990, a 160 foot crane was needed to set this giant yo-yo into action. It then yo-yoed 12 times. **If it took 30 seconds for each yo-yo, how many minutes did it yo-yo?**

Which would you rather be—a yo-yo or a spinning top?
What is the oldest known toy?
Choose your least favorite toy and describe it in such a way that your classmates would want to buy it.

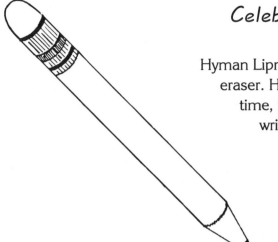

Celebrate the pencil with an eraser!

Hyman Lipman of Philadelphia invented the pencil and attached eraser. He received a patent for his invention in 1858. Until this time, people used separate erasers when completing any writing or drawing. Lipman's pencil offered a groove in which a piece of prepared rubber was glued. The patent was sold for $100,000. You can make about 4,000 check marks with a pencil before you have to sharpen it. **How many pencils could you use to make 12,000 check marks without having to sharpen them?**

CHALLENGE *Why are pencils numbered 2, 2 1/2, or 3?*

Which is easier to erase—a written mistake or an unkind word? Explain why. What is one mistake that you would like to erase?

The Eiffel Tower was built.

Gustave Eiffel, born on December 15, 1832, also designed the steel skeleton for the Statue of Liberty in 1884. The Eiffel Tower, built in 1889, now houses a radio antenna on top. This raised the height of the tower from 300 meters to 320 meters. **How many more feet were added?**

CHALLENGE *Every 7 years, 50 tons of paint are needed to paint the Eiffel Tower. How many times could the Tower have been painted since 1889?*

Compare and contrast the Eiffel Tower to the Statue of Liberty. Which would you rather visit?
The answer is The Eiffel Tower. What could the questions be?
List all the tall buildings, statues and monuments you can think of. Why do you think they were built so tall?

Write a word problem in the ice cream. Write at least 5 sentences. Use the words ice cream, cartons, cones, and Joey in the problem.

No, this is not an April Fool's Day joke!

The leatherback turtle is one of nature's larger animals. These turtles are found in the Pacific Ocean and weigh up to 1,000 pounds. An adult turtle is 6 to 7 feet in length. One turtle, caught off the coast of Monterey, California, in 1961, weighed 1,906 pounds. It was 8'4" long. **How much more does the Monterey turtle weigh than you?**

How is a turtle like a beach ball?
Make a list of names for these gigantic turtles.
What if turtles lived in trees?
Make a mobile of other animals that weigh over 1000 pounds.

April 2

Hans Christian Anderson was born today in 1805.

His father was a shoemaker and made Hans wooden shoes to wear. Hans Christian Anderson wrote over 160 fairy tales. Some of his stories were about his own life. Two of his famous ones were *"The Ugly Duckling"* and *"The Princess and the Pea."* Hans also wrote plays and novels. In 1835, he wrote his first fairy tale. **How old was he?**

List characters that you could write a fairy tale about.

_____   _____   _____

_____   _____   _____

Pick one and write a tale "A Shoe for _____ "
Read "The Ugly Duckling."
How would you feel if you had been the ugly duckling's friend?

_____

_____

_____

_____

_____

The Pony Express began in 1860.

Horseback riders rode as fast as they could to carry U.S. mail. The route, from St. Joseph, Missouri, to Sacramento, California, followed the Oregon-California Trail. Fresh horses were kept at relay stations about 10 to 15 miles apart. The very first Pony Express trip took 10 days and covered 1,966 miles. **If 26 riders went on the first trip, estimate how many miles each rider rode?.**

How is a Pony Express rider like a postcard?
List names that the riders might have called their horses.
Why was it called the Pony Express?
Draw a map (with a legend and key) for a Pony Express rider from your town to the next nearest town.

April 4

Linus Yale, American inventor, was born in 1821.

Linus Yale, Jr., invented the cylinder lock in 1860. He worked with his father to develop locks for banks. Most early locks were not as secure or as small as the Yale lock. Each lock contained a bolt that moved in and out of a slot. The Yale's lock was more complicated and therefore more efficient. **If you had a front door and a back door plus a basement door and a garage door, how many locks would you need to put one on each door?**

Why was the lock an important invention?
Name all the keys that you can.
What other ways can you "lock" something up?
What is a locket?
Draw a picture of something you have "locked" in your heart.

April 5

Tina Maria Stone set a record for a long distance run.

Tina Stone was born in Naples, Italy. When she came to the United States, she moved to California with her family. She became a long distance runner and set a record for running 15,472 miles in one year. **How many miles did she average each week?**

How is a runner like a string bean?
Time some of your friends as they run 25 yards.
Chart the results.

Harry Houdini, a famous magician, was born in 1874.

Mr. Houdini's real name was Ehrich Weiss and he grew up in Wisconsin. His parents came from Budapest, Hungary. He loved to do magic tricks and became famous for his daring underwater escapes. Harry began performing magic tricks when he was 17 years old in New York City. **If Mr. Houdini did 6 magic tricks every 5 minutes for an hour, how many tricks would he do all together?**

How is a magician like a racehorse?
Interview Mr. Houdini after he has completed one of his daring underwater escapes.
Design a poster advertising his magic show.

An enormous chocolate Easter egg was made.

The ostrich and the hummingbird lay nature's largest and smallest eggs. An ostrich egg weighs about 3 1/2 pounds and a hummingbird's egg weighs less than an ounce. This man-made chocolate Easter egg weighed 7,561 pounds and 13 1/2 ounces and was made in England. The egg measured 10 feet high. **Estimate how many ounces more an ostrich egg weighs than a hummingbird egg.**

Name things that weigh between an ostrich egg's weight and a hummingbird egg's weight.
How many children could share this chocolate egg?
Name things that could be carried in an Easter basket.
Draw a design that could be put on an Easter egg.

Hank Aaron broke the home run record in baseball.

Hank broke Babe Ruth's record of 714 home runs. In the days after he broke the record, he received over 900,00 letters. The U.S. Post Office thought this was a record amount of mail for one person to receive. Hank Aaron eventually scored 755 home runs in his lifetime. **How many more home runs did he score after he broke the record?**

Name other baseball players and the records they have broken.
How is a baseball like a hammer?

Design a stamp to honor Hank Aaron.
What baseball player will be the next person to have a stamp designed in his honor? Why?

The United States opens a free public library today.

Libraries become important with the invention of paper and the printing press. Harvard University's library, the earliest in the United States, was begun in 1638. Early libraries all charged their patrons a fee to borrow books. On April 9, 1833, a tax-supported library was opened in Petersborough, New Hampshire. People could now borrow books for free. The idea of the free library quickly spread to other towns. **If Jamestown, the first American colony, was settled in 1607, how many years later did the first library appear?**

How are books like adventures?
How is a library like a peanut butter and jelly sandwich?
Start a class list of all the library books that the students read each month.
Categorize the books and make a graph to show what kinds of books your class reads.
Design a bookmark to celebrate libraries.

This is National Be Kind to Animals Day.

The American Society for the Prevention of Cruelty to Animals (ASPCA) was founded in 1866. The ASPCA founder, Mr. Henry Bergh, was worried about work animals. Today many ASPCA groups have shelters for lost or unwanted pets. ASPCA's want to remind people to be kind to all animals. **If a mother kitten had 8 baby kittens and one fourth of them were adopted by other families, how many kittens would the mother cat have left?**

CHALLENGE What percent of kittens were adopted?

How is a kitten like a milk bottle? List all the names of pets that children in your class have. Sort the pet names into categories of pets and put the information in a graph. Which pet would you like to be? Why? As that pet, which would be your most favorite things to do? least favorite?

A coast to coast walk began in New York City and ended in Los Angeles.

John Lees walked across the United States in 1972. He started today and finished on June 13, 1972. **How many days did it take him to walk from the Atlantic coast to the Pacific coast?**

What other ways could you travel across the country? List the states that Mr. Lees would have walked through if he chose the shortest route from New York to California. Draw a picture of something you might see if you walked across the United States. Write a paragraph that tells about your picture.

Yuri Gagarin was the 1st man in space in 1961.

Men have been interested in traveling in space for many years. In the early 1900's a Russian schoolteacher wrote a paper on how to use rockets to fly into space. No one paid much attention to his scientific writing. About 30 year later, Russia, Germany and the United States began to build space vehicles. Finally in 1961, Yuri, a Russian cosmonaut, was the first man to go into space. He made a single orbit around the earth. Yuri was 27 years old when he traveled into space. **In what year was he born?**

How might the earth look to someone in space? What is a cosmonaut? What things might you find in space? Draw a picture of what you might see if you are looking out a space ship window.

Thomas Jefferson was born in 1743.

Mr. Jefferson, our third U.S. President, was born in Virginia. He went to school when he was five years old. At nine, he was studying Latin, Greek and French. Mr. Jefferson helped write the Declaration of Independence in 1776. He introduced our currency as we know it today. He also wrote books, designed his home at Monticello, and invented the reclining chair. **How old was he when he wrote the Declaration of Independence?**

How is a reclining chair like a merry-go-round? List different types of currency. What might a quarter say to a penny? Design a new piece of paper money to celebrate man's journey into space.

April 14 — A record snowfall fell in Silver Lake, Colorado.

Snow is really ice crystals that form from the water vapor in the clouds. If the drop falls through warm air, it is a raindrop; however, if it falls through cold air it becomes a snowflake. During one 24 hour period, 76 inches of snow fell at Silver Lake, Colorado, on April 14 and 15, 1921. This was a record for the amount of snow in such a short time. **How many feet of snow would this be?**

How is a snowflake like an ice cream cone? Ask an adult if they remember reading about this snow storm. What was the largest snow storm that they remember? that you remember? Write a paragraph about both snowstorms. Design and cut out snowflakes of different geometric shapes. Mindmap manmade or natural objects with the following shapes: diamond, circle, rectangle, square, and triangle.

April 15

McDonald's Restaurant started serving hamburgers in 1955.

Ray A. Kroc, who began the McDonald's chain, served hamburgers from his restaurant outside of Chicago, Illinois. Ten years later, McDonald's had over 4600 restaurants in 23 different countries. These stores have sold approximately 22,000,000,000 hamburgers. **If you ordered a hamburger for $.59 and a cheeseburger for $.69 and French fries for $.89, estimate how much would you spend. Figure out exactly how much you would spend.**

How is a hamburger like a snowstorm? How many McDonald's hamburgers have you eaten? Take a class survey of each person's favorite McDonald's food. Make a chart to show who likes what. Interview a worker at McDonald's about his job.

April 16 — The first books of postage stamps were issued.

Early stamps had no separation lines. Clerks had to use scissors or fold and tear each stamp from a sheet. Later in 1904, punched holes allowed clerks to easily separate the stamps. The first books of stamps in 1900 contained either 12 2¢ cent stamps costing a quarter; or 24 2¢ cent stamps costing 49 cents; or 48 2¢ cent stamps costing 97 cents. **If you bought each one of these books, how much would you spend?**

What was the first stamp issued in the U.S.? What is your favorite stamp? How is a postage stamp like the flag? Start your own collection of stamps. Develop a unique way to present your collection.

The first horses are brought to the colonies.

Matthew Cradock had horses brought to the Massachusetts Bay Colony in 1629. The Chinese calendar celebrates the horse for an entire year. What other animals are celebrated in their calendar? What animal is celebrated this year?

Did you know that Christopher Columbus brought horses with him on his second voyage to the new world? What role did the horse play in the exploration of the United States? Make a timeline of events linked with the horse. Create a new animal to be celebrated during one Chinese year. When was the calendar first invented? What if we didn't have calendars?

Pringle's produced its largest potato chip.

The Pringle chip, made in Tennessee in 1990, was 23 inches by 14 1/2 inches. The first potato chips were made by Earl Wise in the 1920's. He had too many potatoes in his restaurant so he peeled and sliced them with a cabbage cutter and then fried them. He sold them in brown paper bags. **If Mr. Wise sold 100 bags of chips on Monday, 65 bags on Tuesday, 75 bags on Wednesday, 45 bags on Thursday, and 90 bags on Friday, how many bags did he sell during that week?**

CHALLENGE How many bags of chips did Mr. Wise average for the week?

What other uses can you think of for a potato? Which do you like more—potato chips or pretzels? Invent a new snack food. How would it taste, smell and feel?

The first kindergarten for the blind was dedicated in 1887.

The school opened on May 2, 1887, in Massachusetts with 10 children. Today, over one-half of the blind children in the United States go to a regular school. Many also go to college. **If twice as many blind children joined the first kindergarten the following year, how many entered school?**

How would you describe the color red to a blind person? purple? yellow? Learn to read braille. Teach a friend. Try writing a note to a friend in braille. What is your most important sense? Why?

Expo '92 opened in Spain.

Over 100 countries celebrated *"The Age of Discoveries"* at this World's Fair in 1992. The 1939 World's Fair held on Long Island, New York, opened with a special exhibit, *Futurama*. People sat in chairs and moved over the landscape of the future. Planners of *Futurama* predicted cars with air conditioners and throw away houses. **If the admission cost of the 1939 World's Fair was 25 cents, how many different ways could you make change from a $5.00 bill?**

What would you find in a Futurama exhibit at the next World's Fair? When and where is the next World's Fair? Create a new exhibit for this fair.

April 21

The first firehouse pole was installed.

Captain David Kenyon of Engine Company No. 21 in New York City installed the pole in 1878. A hole was cut in the upper floor and a 3 inch wide greased pole extended to the floor below. This helped firemen reach the first floor more quickly than using the stairs. **If it took a fireman 1/2 minute to reach his truck by stairs and 8 seconds using the pole, how much time was saved?**

What other objects help us save time?
What other things are saved?
Design a piggy bank that would help you save money.

April 22

Coins are stamped with a saying.

"In God We Trust" was first stamped on coins in 1864. Did you know that coin collecting was known as the "hobby of the Kings?" Only kings and rich people were collectors at first. Today, this hobby is shared by over 5 million numismatists or coin collectors. Coins have been minted in half cents to $20 gold pieces. Only six coin denominations are now being minted. **What are they?** Measure the diameters of as many of these coins as you can find. **What is the difference between the coin with the largest diameter and the coin with the smallest diameter?**

What would a coin from 1864 tell you? A new coin is designed after many sketches. A large model of the coin is then made. It is carved into a smaller form to make the size of the coin as we know it. Sketch a design for a new coin that would represent your class. Take a vote on the best coin.

Celebrate the birthday of Granville T. Woods.

Granville Woods, a black inventor with over 500 patents, designed a telegraph in 1855 which let dispatchers communicate with moving trains. Because of his invention, less trains collided with each other on the tracks. He also invented an electric incubator used for hatching chicken eggs. **If 24 eggs were put into an incubator and only 18 hatched, how many did not?**

CHALLENGE What percent of the eggs hatched?

What other animals are hatched from eggs? List other things that are hatched. List other inventions that help save people's lives like the telegraph. Which one is most important? Explain why.

The world's oldest clown was born.

Charlie Revel began performing in 1899 and ended his clowning career in 1981. He was known as the world's oldest clown. In Venice, Florida, there is a school called Ringling Brothers and Barnum and Bailey Clown College. You have to be 17 years old to apply. It takes just 9 weeks to graduate as a clown. **If Charlie Revel was born in 1896, how many years was he a clown?**

Why are clowns important to a circus? What things make you laugh? cry? What act would you like to perform in a circus? Draw a picture of you performing in the circus.

A seeing eye dog was presented to his owner.

Buddy, a shepherd from Switzerland, was presented to his blind owner, Morris Frank, in 1928. He was the first seeing eye dog. The dogs are trained in special schools to work with blind people. The first such

school for these dogs in the United States was "Seeing Eye, Incorporated." It was established in 1929. It takes four weeks for dogs to become friends with their blind owner. **How many days is this?**

Seeing eye dogs are also known as guide dogs. Why? In what other ways do animals help us? Which animal is the most important? Why? If possible, invite a blind person with a seeing eye dog to class. How does the owner communicate to the dog? How does the dog communicate with the owner? Develop a list of questions you could ask.

John Audubon, artist and naturalist, was born in 1785.

John Audubon was well known for his drawings of birds in their habitats or surroundings. He was famous for his colored engravings of 435 life-sized birds in their natural habitats in the book **Birds of America.** The National Audubon Society was named after him. Their society's mission is to protect birds and their environments. **How old was Audubon when his book was published in 1826?**

How would the world be different if there were no birds? Which would you rather be—a pelican or a penguin? How is a bird like a kite?

The first yellow baseball was used.

This regulation National League baseball was used in the Columbia-Fordham game in New York City in 1938. It was dyed yellow with red stitches and developed by Frederic Rahr. A baseball player may have a batting average of .353. This represents the probability that he will get a hit. **What is the chance he will not get a hit?**

CHALLENGE If a batter hits a home run 40% of the time she is at bat, how many home runs will she hit if she is at bat 10 times during one game?

Write a poem that helps describe the definition of probability. Illustrate your poem. Draw a picture about something that will probably happen to you tomorrow.

Celebrate Great Poetry Reading Day!

Poetry expresses feelings or imagination. The longest poem ever written in English was 129,807 lines long. It was written about King Alfred and took John Fitchett over 40 years to write. **Write the numeral 129,807 in words.**

CHALLENGE If each day Mr. Fitchett wrote an equal number of lines, approximately how many lines did he write each day?

How is a poem like a circle? Write a poem or a story about any shape. Illustrate your poem. Write a pun, riddle or joke about any shape.

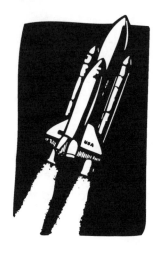

The Challenger space shuttle lifts off.

This crew had seven members. Its was the first recovery and re-pair mission of a satellite that was not working. A robotic arm helped pull the satellite from space so that the astronauts could fix it. The satellite was then re-launched into space. The shuttle made 111 orbits around the Earth and returned on May 6, 1985. **How many days was the Challenger in space?**

CHALLENGE Estimate how many orbits around the earth were made each day.

Name different forms of transportation.

_____   _____   _____

_____   _____   _____

Find a unique way to categorize them.

A_____   B_____   C_____

_____   _____   _____

_____   _____   _____

Which would you rather ride—a supersonic jet, a monorail or a catamaran? Why?
I'd rather ride a _____ *because* _____

_____

The Betsy Ross Bridge was opened in 1976.

This was the first bridge named for a woman. It spanned 3 miles, connecting Philadelphia and Pennsauken, New Jersey. Lake Pontchartrain Causeway # II, a combination bridge is 126,055 feet long. **How much longer is it than the Betsy Ross Bridge?**

Make a chart showing the different kinds of bridges. Compare and contrast a bridge with a straw. The U.S. Flag flies over the grave of Betsy Ross at Mt. Moriah Cemetery in Pennsylvania. This is one of the few places the flag flies at night. Research other places the flag flies at night as a patriotic gesture. Which place would you like to visit most? Write a travel article about the importance of this place.

How many word problems can you make about Nancy and her Lemonade Stand? You can include other kids, selling lemonade, and Nancy's dreams about becoming rich!

Today is Mother Goose Day. May 1

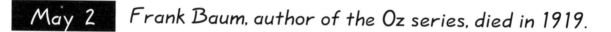

Although Mother Goose was not real, we give her credit for many nursery stories and rhymes that are favorites with children. The nursery rhymes were probably passed down from family to family the same way that fairy tales were. In 1697, a Frenchman named Charles Perrault wrote down these rhymes and published a book titled *Tales of Mother Goose.* Since then many authors have published Mother Goose rhymes. Some favorite rhymes are *"Little Jack Horner"* and *"Hey Diddle Diddle."* Research Mother Goose rhymes, and list ones that have numbers in them.

How is Mother Goose like Hans Christian Anderson? Write your favorite rhyme and make a class collection of all the class favorites. Memorize your favorite rhyme and perform it for your classmates using puppets or masks.

May 2 Frank Baum, author of the Oz series, died in 1919.

Mr. Baum, born on May 15, 1856, wrote 14 children's stories about the land of Oz. One of the fourteen Oz books, **The Wonderful Wizard of Oz,** was made into a movie in 1939. In the story, Dorothy and her dog, Toto, arrive in the land of Oz in a very unusual way. In Oz, she meets some unusual characters including *munchkins*. Even though she enjoys her new friends, she wants to return home to Kansas. Read the story and discover how her special ruby slippers help her to return home. **How old was Mr. Baum when he died?**

CHALLENGE What percent of Mr. Baum's Oz stories became movies?

What questions would you ask a munchkin? Would you rather live in the Emerald City or on the Ruby Ranch? Why? Draw a picture showing how you think the Land of Oz might look.

Get ready! The circus is coming to town! May 3

Five brothers, Albert, Otto, Alfred, Charles and John, formed the famous Ringling Brother's Circus. On May 19, 1884, the brothers performed in Baraboo, Wisconsin. Along with 17 other employees, they sewed and pitched the tent, sold tickets, played in the band, and performed all the circus acts. At first the

circus traveled in wagons from town to town. As they grew and became more popular, they were able to afford to travel by train. In 1907, the Ringling Brothers and the Barnum Bailey circuses joined and became the Ringling Brothers Barnum Bailey Circus. The Ringling family sold the circus in 1967. The circus's name has stayed the same. **How many years was the Ringling family in the circus business?**

List acts that might be in a circus. Name things that come in rings. How would you feel if you were the trapeze artist in a circus? the trapeze?

Mrs. Delina Filkins was born today in 1815 in Stark, New York. **May 4**

When Mrs. Filkins died on December 4, 1928, she had lived longer than almost anyone else in the world. Some animals live longer than humans, but most live a very short time compared to you and me. On the average, elephants live 35 years; camels live 12 years; and pigs live 10 years. **How many years did Mrs. Filkins live?**

What would a school day be like in 1815? Name things that were invented during Mrs. Filkin's lifetime. Name things that have been invented in your lifetime. Create a crossword puzzle with the names of the inventions.

May 5

Alan Shepard, astronaut, was the first American in space.

The first trip into space was made by a Russian space craft called Sputnik I. Then Russians and the United States began racing to see which country could find out the most about outer space. The first American in space, Alan Shepard, lifted off from Florida in a space capsule and rocketed 115 miles high. His trip lasted only 15 minutes before he landed in the Atlantic Ocean. He then became an Apollo 14 crew member in 1971 and was the 5th person to walk on the moon. **If Mr. Shepard was born in 1923 and was 38 years old when he went into space, in what year would his space trip have taken place?**

How is a space capsule like a tennis ball? What might a bird say to a space ship? Write a poem about what an astronaut might say about his trip into space. Draw a timeline of the United States's vehicles that have gone into space.

Robert E. Peary, Arctic explorer, is born in 1856. **May 6**

Mr. Peary made many trips to the Arctic north before he finally reached the actual position that is called the North Pole. In 1886, Mr. Peary made a trip to Greenland in the first of his efforts to discover the North Pole. He sailed on special ships that were built to sail among the icebergs and then took dog sleds to travel across the frozen ice. He finally reached his goal with his assistant Matthew Henson on April 6, 1909. **How long after he made his first trip to Greenland did he reach the North Pole? How old was he when finally reached the North Pole?**

Write a journal entry for Peary on the day he reached the North Pole. How is the North Pole like a giraffe? Design a new vehicle that can move through ice. Make a time line that includes important dates in the lives of Delina Filkins, Alan Shepard, and Robert Peary.

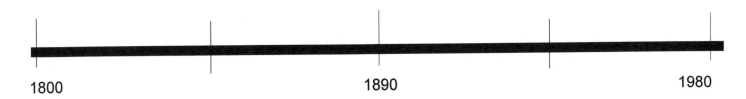

1800   1890   1980

In 1975, a group of men make a record dog sled journey.

Antarctica is like a giant iceberg because much of the land is covered with ice. There are very few plants or animals that live there. Penguins, birds that don't fly, are probably the best known animals that live in Antarctica. The United States set up a scientific base at the South Pole called the Amundsen-Scott Station named after the explorers who had found the South Pole. In 1975, ten people began a 3000 mile dog sled trip across Antarctica. It took them 300 days. **How many miles a day did they travel?**

What kinds of things might they see on their journey in the Antarctic?
What things would they need to take along on their trip to survive?
Compare and contrast a dog sled and an ice cube.
What kinds of animals live in cold countries?
Write a letter to the editor of the newspaper telling why dogs should remain in the Antarctic.

May 8

Today a large goldfish was caught in Texas.

There are over 30,000 different kinds of fish that live in oceans, lakes and streams. Some fish live about one year while others live over 50 years. In 1988, a man caught a goldfish that weighed 3 1/2 pounds and was 15 1/4 inches long. He used homemade dough on his hook. **Estimate how big a fish tank would need to be to hold this goldfish. Name things that weigh less than 3 pounds.**

If you could interview this goldfish, what questions would you ask?
What kinds of things might you use to catch a goldfish?
Would you rather catch a goldfish or a shark? Why?
Invent a new way to catch goldfish.

Christopher Columbus and Ferdinand, his son, set sail for Central America.

There are many stories about Christopher Columbus's trips to America. His first voyage was in 1492 with three ships. On his second trip he commanded 17 ships. On this trip many men sailed with Columbus because they wanted to live in the "new" world. On his fourth trip to the new world, Christopher Columbus took his 13 year old son, Ferdinand, with him. Many of the sailors on the ship were young boys about Ferdinand's age. **If Columbus and his son spent 30 months in Central America and 6 months sailing back to Spain, how old would the son be when they reached home?**

What would you ask Christopher Columbus if you met him today?
What would you ask his son?
What would Ferdinand say to you if he visited your town and school?
Write a letter to persuade Mr. Columbus to take you along on one of his journeys.

The first railroad built across the western United States was completed in 1869.

Railroad became a popular means of transportation in the 1800's. In 1869, there was a railroad from the Atlantic Coast to the Pacific. The last section of this railroad was the Union Pacific-Central Line. It was 1,780 miles long and went from the Missouri River to the Pacific Ocean. It was one of four railroads to be built in the 25 years after the Civil War. **If you boarded the train in Missouri and it took you 10 days to travel to the end of the line, how many miles would you travel a day on the average?**

Name things that take you from one place to another.
Invent a new way to travel across the desert.
What would a speeding train look like to a flower growing along the track?
Write a poem about a train engine.

In 1926, Commander Richard E. Byrd flies over the North Pole.

Byrd was another explorer like Robert E. Peary who discovered the North Pole. He first flew over Greenland in 1925. In an historic flight, Navy Admiral Richard E. Byrd and his co-pilot flew over the North Pole. Several years later, he flew over the South Pole in Antarctica. Admiral Byrd also spent one winter living at a base in Antarctica and wrote a book about it. He flew over the pole two more times before his death. **If Byrd lived 31 years after he first flew over the North Pole, in what year did he die?**

How is an airplane like a pizza?
Living at one of the poles is like camping out in winter.
Write a week's worth of journal entries pretending you are living at a camp at one of the poles.
How would your life be different if you had no electric lights at home or in school?

May 12

Captain Joshua Slocum sailed around the world on Spray.

In 1895, Captain Slocum left Boston to sail around the world. His boat called *Spray* was only 36 feet long. He returned to the United States in 1898 after sailing more than 46,000 miles. He was the first man to sail around the world alone. **How long did it take him?**

On an atlas, follow Captain Slocum's journey around the world, and list the bodies of water where he might have sailed.
What would you need to take with you on an "around the world trip" by sailboat?
How is a sailboat like a jigsaw puzzle?

A Pennsylvania man jump-roped his way to a record in 1989.

In one hour, Mr. Robert Commers jumped 13,783 times without missing. His old record was 623 jumps. **How many more times did he jump to set the second record?**

Name things that jump. How do you feel when you set a new record for something? Take turns with your classmates and each of you jump rope for 3 minutes. Record the number of times you each can jump without missing. Make a graph of the classes' jumps. What other types of records could your class attempt? (Remember safety first!)

Alaska is bought by the United States.

Alaska is the largest peninsula in the Western part of the world. The United States paid Russia about $7.2 million which was about two cents an acre for Alaska. People thought that this was too much money to pay for this cold land near the Arctic Circle. Today Alaska is our largest state but has the fewest number of people living there. It supplies us with oil, fish and lumber. **If Alaska is 50 miles from the Soviet Union, how far away is it in kilometers?**

What state has the most people living in it? What things can you buy for two cents? How is a penny like a fish? Interview a real estate agent and ask how much land costs per acre today. Design a brochure to encourage people to move to Alaska.

The rare white Steiff bear was sold for a record amount in Germany.

This very expensive teddy bear was sold for $15,840 in 1987. The most expensive bear sold before was at Sotheby's in London, England, and it cost $9,152. **How much more expensive was the white bear?**

There are 8 different kinds of real bears. Make a chart showing how they are alike and how they are different. List names for teddy bears. How is a teddy bear like a panda bear? If you could interview the white Steiff bear after it was sold, what questions might you ask it? Draw a picture of the bear, and show how it might have felt after it was sold.

The nickel or five cent piece was first used in 1866.

The nickel was made of 75% copper and 25% nickel. It had a shield on one side and a numeral 5 on the reverse. What other pictures have been on the nickel? **If you tossed a nickel 50 times, how many heads would you expect?** Try it yourself. See if the results are what you predicted.

Which president's face is stamped on a nickel? List all the ways you can use a nickel other than as money. What can a nickel buy? Start your own coin collection or visit a collection started by a friend or adult. Why is an "Indian Head" nickel so valuable?

May 16

"Sugar" Ray Leonard, boxer, was born in 1956.

May 17

Mr. Ray Charles Leonard won many boxing titles. He also won a gold medal in the 1976 Olympic Summer Games. He was named "Sugar" after another boxing champ, Sugar Ray Robinson. **If Sugar Ray won 8 out of the past 10 boxing matches, what is the probability that he will win the next match?**

List occupations of people who make a living using their hands. Why are their hands important to their jobs?

Categorize these occupations.

Unskilled labor

Skilled labor

Professional

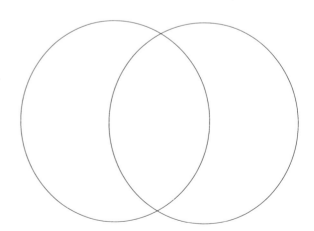

Using a Venn diagram, find similarities and differences between two of the occupations.